Annals of Mathematics Studies

Number 196

Hölder Continuous Euler Flows in Three Dimensions with Compact Support in Time

Philip Isett

PRINCETON UNIVERSITY PRESS
PRINCETON AND OXFORD
2017

Published by Princeton University Press, 41 William Street, Princeton, New Jersey 08540

In the United Kingdom: Princeton University Press, 6 Oxford Street, Woodstock, Oxfordshire
OX20 1TR
press.princeton.edu

Library of Congress Cataloging-in-Publication Data

Names: Isett, Philip, 1986–
Title: Hölder continuous Euler flows in three dimensions with compact
    support in time / Philip Isett.
Description: Princeton : Princeton University Press, [2017] | Series: Annals
    of mathematics studies ; number 196 | Author's thesis
    (doctoral)—Princeton University, Princeton, N.J., 2013 | Includes
    bibliographical references and index.
Identifiers: LCCN 2016042604| ISBN 9780691174822 (hardcover : alk.
    paper) | ISBN 9780691174839 (pbk. : alk. paper)
Subjects: LCSH: Fluid dynamics—Mathematics.
Classification: LCC QA911 .I84 2017 | DDC 532/.05dc23
LC record available at https://lccn.loc.gov/2016042604

British Library Cataloging-in-Publication Data is available

The publisher would like to acknowledge the author of this volume for
providing the camera-ready copy from which this book was printed.

This book has been composed in LaTeX

Printed on acid-free paper ∞

10 9 8 7 6 5 4 3 2 1

Contents

Preface

The present monograph concerns the construction of Hölder continuous weak solutions to the incompressible Euler equations that fail to conserve energy. This manuscript originally appeared as an arXiv preprint in November 2012, and evolved into the author's Ph.D. thesis in the Princeton math department [Ise13b]. The greatest motivation for constructing such solutions has come from the pursuit of Onsager's conjecture, which states that for any $\alpha < 1/3$ there exist weak solutions to incompressible Euler with Hölder exponent α for which the total kinetic energy fails to be conserved. In the following we show the existence of $(1/5 - \epsilon)$-Hölder Euler flows that fail to conserve energy, which is at this time the best result towards proving Onsager's conjecture in its original formulation.

A few years have passed since the first appearance of this work, and the study of weak solutions to incompressible Euler and related equations has seen many developments in this time. To provide an account closer to the state of the art of this still very active field of research, we provide here a list of some references to recent work and comment briefly on the relationship to the present work. We will present some more detailed discussion in the Appendix Section (B).

Shortly following the appearance of the present work, a second construction of $(1/5 - \epsilon)$-Hölder Euler flows was given by Buckmaster, De Lellis and Székelyhidi [BDLS13]. Their main result shows also that the kinetic energy profile may be prescribed to be any strictly positive smooth function. The proof given in [BDLS13] is much shorter than the one given here, and adheres more closely to the previous work of [DLS13, DLS14]. We refer to [BDLIS15] for a self-contained presentation of the proof in [BDLS13] that includes a more detailed technical comparison to the present work.

In [Buc15, BDLS15], Buckmaster, De Lellis and Székelyhidi have constructed Euler flows with Onsager critical spatial regularity in the class $C^0_{t,x} \cap L^1_t C^{1/3-\epsilon}_x$. We note that this result has required some technical departures from the construction given here. We also note that the argument of [BDLS15] makes crucial use of the technique of "mollification along the flow" introduced below, whereas in [BDLS13] this technique had been replaced by a simpler alternative.

In [Ise13a], the author presented a theory of time regularity for Euler flows in the class $L^\infty_t C^\alpha_x$, and proposed a stronger form of Onsager's conjecture called the "Generic Failure of Energy Regularity," which is elaborated in [IOa]. An extension of the present construction to produce solutions with compact support in space and time is given in [IOa, IOb], which contain also some first steps towards proving the conjectured "Generic Failure of Energy Regularity" for solutions with Hölder exponent $\alpha < 1/5$. A version of the present construction has also been extended to a general class of active scalar equations in [IV15], where an analogue of Onsager's conjecture is expected to hold. The works [IOa, IOb] and [IV15] demonstrate the robustness of the organizational approach of the present book by showing how the same Main Lemma can be used to address questions concerning (1) h-principles, (2) nonuniqueness of solutions to the initial value problem, (3) prescribing of a given energy profile, (4) failure of higher regularity of solutions and (5) existence of compactly supported solutions. These works also contain several techniques that can be used to simplify the argument given below.

Hölder Continuous Euler Flows in Three Dimensions with Compact Support in Time

Part I
Introduction

In the paper [DLS13], De Lellis and Székelyhidi introduce a method for constructing periodic weak solutions to the incompressible Euler equations

$$\begin{cases} \partial_t v + \operatorname{div} v \otimes v + \nabla p = 0 \\ \operatorname{div} v = 0 \end{cases} \tag{1}$$

in three spatial dimensions that are continuous but do not conserve energy. The motivation for constructing such solutions comes from a conjecture of Lars Onsager [Ons49] on the theory of turbulence in an ideal fluid. In the modern language of PDE, Onsager's conjecture can be translated as follows.

Conjecture (Onsager (1949)).

1. Weak solutions to the incompressible Euler equations that obey a Hölder estimate

$$|v(t, x + y) - v(t, x)| \le C|y|^\alpha$$

   for some $\alpha > 1/3$ must conserve energy.

2. Furthermore, for any $\alpha < 1/3$, there exist weak solutions to the Euler equations that belong to C^α and fail to conserve energy.

Onsager's conjecture can be appreciated in the context of the theory of turbulence famously proposed by Kolmogorov [Kol41] in 1941. One key postulate of Kolmogorov's theory is an anomalous dissipation of energy for solutions of the three-dimensional Navier-Stokes equations

$$\begin{cases} \partial_t v + \operatorname{div} v \otimes v + \nabla p = \nu \Delta v \\ \operatorname{div} v = 0 \end{cases} \tag{2}$$

in the low viscosity regime $\nu \to 0$ (or, more precisely, at high Reynolds number). One formulation of anomalous dissipation is that a sequence of solutions v_ν to the three-dimensional Navier-Stokes equations with the same initial data $v_\nu(0, x) = v(0, x)$ may have energy functions $e_\nu(t) = \frac{1}{2} \int |v_\nu|^2 dx$ that do not converge to a constant function of time as $\nu \to 0$, but rather may possess some energy dissipation independent of the viscosity parameter. Kolmogorov's theory proposes that the phenomenon of anomalous dissipation is generic in a statistical sense for ensembles of solutions to the Navier-Stokes equations at low viscosity. The limiting energy dissipation rate is one of the main quantities that are proposed to govern the statistical properties of turbulent flows in Kolmogorov's theory. We refer to [Fri95] for a more detailed account.

Onsager proposed that one may be able to observe anomalous dissipation of energy even in the absence of viscosity in the context of the Euler equations (the case $\nu = 0$ of Navier-Stokes), but pointed out that only low regularity solutions to the Euler equations can exhibit turbulent behavior since smooth solutions must conserve energy. By studying the interactions between different frequency components of the solution that arise from the nonlinearity, Onsager proposed that anomalous dissipation could be explained in terms of a transfer of energy from coarser to smaller scales, and deduced that the exponent $1/3$ should be critical for energy conservation. His notion of solution was based on a Fourier series representation, but it can be shown to be equivalent to the modern notion of a weak solution. A review of his computations can be found in the note [DLSa].

It is known that solutions to the incompressible Euler equations with Hölder regularity greater than $1/3$ conserve energy in any dimension, so part (2) of Onsager's conjecture has been settled. A short proof of this statement was presented in [CET94] after a slightly weaker result was established in a series of papers by Eyink [Eyi94] following Onsager's original computations. We provide a proof in the Appendix Section (A) for the interested reader. More precise results, as well as a discussion of what "Onsager critical" function space could best be used to model ideal turbulence, can be found in [Shv10]. In recent years, substantial progress has also been made toward constructing dissipative solutions with Hölder regularity less than $1/3$.

The first proof that weak solutions to the Euler equation need not conserve energy came in a groundbreaking paper of Scheffer [Sch93], in which he produced weak solutions to the Euler equations with compact support in space and time belonging to the class $L^2(\mathbb{R}^2 \times \mathbb{R})$. Following Scheffer's discovery, in [Shn97], Shnirelman found a simpler construction of weak solutions in the class $L^2(\mathbb{T}^2 \times \mathbb{R})$ with compact support in time. Shnirelman later in [Shn00] produced weak solutions in the class $L_t^\infty L_x^2(\mathbb{R} \times \mathbb{T}^3)$ which dissipate energy using the concept of a generalized flow introduced by Y. Brenier.

In the breakthrough paper [DLS09], De Lellis and Székelyhidi were able to construct weak solutions in the class $L^\infty(\mathbb{R}^n \times \mathbb{R})$ for any $n \geq 2$. In a subsequent paper [DLS10], they were also able to produce solutions belonging to the energy space $C_t L_x^2$, and their main theorem demonstrates that the energy density of these weak solutions, $\frac{1}{2}|v|^2(t, x)$, can be set equal to essentially any prescribed non-negative, continuous function $e(t, x)$. These breakthroughs led to new results concerning weak solutions to several equations of fluid dynamics, which are surveyed in [DLSb], and also demonstrated that many entropy criteria one might propose are unable to recover uniqueness of solutions in the energy class. The constructions are performed through a technique known as convex integration, which originated in the work of Nash on C^1 isometric embeddings [Nas54] and was extended by Gromov to become a tool for establishing the h-principle in many other applications to topology and geometry (see [Gro86]). For the background of the reader, we review the history and main ideas of this technique. We then summarize the newer developments made in the context of the Euler equations, and the contributions of this monograph.

The Technique of Convex Integration

When solving differential equations, the essence of the convex integration procedure, beginning with the work of Nash, is to first formulate a notion of "subsolution" to the equation one is trying to solve, and then to show that any given subsolution can be altered by adding a sequence of highly oscillatory corrections in such a way that a solution is achieved in the limit. For isometric embeddings of, say, $u : S^2 \to \mathbb{R}^3$, Nash's notion of a subsolution is that of a short map, that is, a smooth map $u_0 : S^2 \to \mathbb{R}^3$ such that the pullback of the Euclidean metric $Du^T Du$ is pointwise less than or equal to the metric on the sphere as a quadratic form; equivalently, a short map is one for which the arclength of $u(\gamma)$ is less than the arclength of γ for any curve γ on S^2. For example, rescaling the standard sphere into a smaller ball is a short map. Clearly, any map which can be uniformly approximated by isometric embeddings will be a short map. To obtain an isometric embedding approximating an initial short map u_0, one adds a sequence of oscillatory corrections to u_0 which can be chosen arbitrarily smaller and smaller in the C^0 norm, but which still make significant changes in the derivative Du. Iteratively adding such oscillations, one obtains a sequence of short maps converging in C^1 to an isometric embedding in the limit[1]. On the other hand, the C^2 norms of these corrections grow without bound, and the embedding obtained by this procedure cannot be C^2, as C^2 isometric embeddings of $S^2 \hookrightarrow \mathbb{R}^3$ are unique up to a rigid motion. We refer to [CDLS12b] for more regarding the h-principle for low regularity isometric embeddings.

In fact, there is a very useful analogy between the isometric embedding problem and the Euler equations where the velocity field v plays a role analogous to the derivative Du of the embedding. Most of the analogies extend from the fact that the nonlinearities $v \otimes v$ and $DuDu^T$ for the two equations are both symmetric, quadratic, and non-negative in the appropriate sense. This analogy was first discussed by De Lellis, Székelyhidi and Conti in their work [CDLS12b] on proving the h-principle for $C^{1,\alpha}$ isometric embeddings. See [DLSb] for further discussion.

The version of convex integration employed in [DLS09, DLS10] for the Euler equations is very different from the original version of convex integration applied to the isometric embedding problem by Nash and extended

---

[1] Actually the "spiral" corrections used in Nash's original argument would require that the u map into a higher dimensional space such as $\mathbb{R}^4$, but shortly after Nash's paper, Kuiper was able to design corrections which can achieve the same goal in the codimension 1 case.

in [CDLS12b]. It is based on an extension of the convex integration scheme used by Müller and Šverák [MŠ56, KMŠ03] to construct solutions to differential inclusions $\nabla u \in \mathcal{R}$ that are only Lipschitz (i.e., $\nabla u \in L^\infty$), rather than C^1 (i.e., $\nabla u \in C^0$). As originally explained by Kirchheim [Kir03], this "weak" version of convex integration for Lipschitz maps can be implemented in an elegant and simple manner through Baire category arguments, or by an "explicit" iteration that is basically equivalent to the proof of the Baire category theorem.

The weak version of convex integration is unable to produce continuous solutions to the Euler equations. Rather, the solutions produced by De Lellis and Székelyhidi in [DLS10] generically have no better regularity than the generic vector field of prescribed energy density $\frac{|v|^2}{2}(t, x) = e(t, x)$ when the set of such vector fields is equipped with a weak topology (for instance, the L^∞ weak-$*$ topology or the topology $C_t H_x^{-1}$; the latter space has the advantage of ensuring that solutions belong to the energy space $C_t L_x^2$). The fundamental obstruction to achieving continuous weak solutions by this variant of convex integration is that even though one can choose the frequencies of the oscillatory corrections so large that the corrections may be arbitrarily small in a weak topology, these corrections are still required to have a certain size in a strong topology (C^0 for Euler) in order for any noticeable progress towards achieving a solution to be measured. The many solutions obtained by this construction are connected to Gromov's h-principle in that one actually shows that the subsolutions used to perform the construction can be approximated (in a weak topology) by solutions, and that the solutions generated by the process exhibit a huge amount of flexibility despite solving the equation.

New Developments

Recently in [DLS13], De Lellis and Székelyhidi have made another outstanding breakthrough by constructing continuous weak solutions to the Euler equation on a periodic domain $\mathbb{R} \times (\mathbb{R}/\mathbb{Z})^n$ ($n = 3$) whose energy $\int_{\mathbb{T}^n} \frac{|v|^2}{2}(t, x)dx$ can be any smooth function $e : \mathbb{R} \to \mathbb{R}_{>0}$ that is bounded below by a strictly positive constant. In particular, these solutions may dissipate energy. They also achieved the same result in $n = 2$ spatial dimensions in the preprint [CDLS12a] with A. Choffrut.

Following their construction of continuous solutions and building on the methods in [CDLS12b], they extended their method in the paper [DLS14] to construct weak solutions to the Euler equations on $\mathbb{R} \times (\mathbb{R}/\mathbb{Z})^3$ with velocity in the Hölder space $C_{t,x}^{1/10-\epsilon}$ and having any prescribed energy obeying the same restrictions as in the continuous case. This result was generalized to the case of two spatial dimensions with the same Hölder exponent of $1/10 - \epsilon$ by A. Choffrut in [Cho13]. The paper [Cho13] also contains more detailed results describing the flexibility of the family of solutions produced by the method.

The convex integration scheme used in these recent results more directly resembles the original scheme used by Nash to construct C^1 isometric embeddings, and it also bears more resemblance to the argument of Shnirelman in [Shn97] than does the argument in [DLS09]. To achieve their results, De Lellis and Székelyhidi have introduced several important, new ideas which represent dramatic changes in the point of view of the convex integration scheme. Although these ideas cannot be summarized at this stage of the introduction, we will refer to them as the analogous aspects arise in the present book, and we urge the reader to study their papers. For now we mention three new aspects:

- They introduce an (underdetermined) system of PDEs called the "Euler-Reynolds equations," which form the correct space of "subsolutions" in which to perform the convex integration procedure.

- They identify the two main error terms in the construction, which they call the "transport part" of the error and the "oscillatory part" of the error. The main difficulty in the procedure is to construct high frequency waves which achieve smallness for both error terms simultaneously.

- Their main idea for controlling the oscillatory part of the error involves the use of a linear space of high frequency, stationary solutions of the Euler equations called "Beltrami flows."

The idea that turbulent Euler flows may be constructed from Beltrami solutions has appeared in the turbulence literature, and was suggested to De Lellis and Székelyhidi by Peter Constantin [DLS13].

In this book, we build upon and rework the convex integration scheme of De Lellis and Székelyhidi in order to achieve the following theorem.

Theorem 0.1. *For every $\delta > 0$, there exists a nontrivial weak solution*

$$v(t,x) : \mathbb{R} \times \mathbb{T}^3 \to \mathbb{R}^3 \tag{3}$$

$$p(t,x) : \mathbb{R} \times \mathbb{T}^3 \to \mathbb{R} \tag{4}$$

to the Euler equations that belongs to the Hölder class

$$v \in C_{t,x}^{1/5-\delta} \tag{5}$$

$$p \in C_{t,x}^{2(1/5-\delta)} \tag{6}$$

such that the support of (v,p) is contained in a compact time interval.

The framework we develop appears robust enough to obtain the regularity of $1/3 - \delta$ conjectured by Onsager except for one term where stationary flows are used in a crucial way. We discuss this difficulty as it arises in the argument. We propose as a conjecture an "Ideal Case" scenario which summarizes what the method would yield if the C^0 norm of this term were suitably well-controlled. This conjecture, if true, could be used to construct energy-dissipative solutions in the Hölder class $v \in C_{t,x}^{1/3-\delta}$, and in particular would imply Onsager's conjecture.

The proof of Theorem (0.1), which builds heavily upon the ideas of De Lellis and Székelyhidi, also implements the method of convex integration. The argument to be presented here is based on their approach in [DLS13], but includes several novel features:

- We use nonlinear phase functions to form the basic building blocks of the construction. This idea provides a more effective means of controlling the Transport term in the error. To implement it, we adapt the method of De Lellis and Székelyhidi for obtaining small solutions to the relevant elliptic equation based on nonstationary phase, and we develop a new method for constructing the amplitudes of the basic building blocks.

- Our construction is organized around a flexible Main Lemma, which summarizes the overall result of a single iteration of the scheme. This Lemma has the feature that one controls the time interval supporting the solutions and the error, allowing us to obtain solutions with support in a finite time interval, and also to prove Theorem (0.2) on the gluing of solutions. This latter result implies nonuniqueness in the class $C_{t,x}^{1/5-\delta}$ for the initial value problem with arbitrary smooth initial data.

- We introduce a sharp, general framework for calculating the regularity achieved by the construction. This framework reduces regularity computations and bounds for other physical quantities to simple, linear algebra calculations.

- Our framework for measuring regularity is based on a notion of "frequency and energy levels" used to measure the size of the error, the approximate solutions, and their derivatives during the iteration process. This notion has the important feature that it distinguishes the bounds for the derivatives of the velocity, the pressure and the error. The frequency energy levels also keep track of second derivative bounds which play an important role in some of the estimates and appear to be necessary for estimating one of the error terms in the conjectural ideal case scenario.

- We isolate the material derivative $\partial_t + v \cdot \nabla$ as a special derivative in the construction. It appears that unless improved bounds for the material derivative are taken into account, the highest regularity one can achieve through this construction is $1/(3 + \sqrt{8}) - \delta$.

To take advantage of the special role of the material derivative, we introduce several additional ideas into the scheme, for example:

6

- We incorporate improved bounds for $\partial_t + v \cdot \nabla$ into the notion of frequency energy levels. In particular, the material derivative obeys better bounds than do the spatial derivatives or the time derivative.

- We use time averaging along the coarse scale flow of the fluid as a special form of mollification.

- We introduce a "Transport-Elliptic" equation in order to eliminate the error in our parametrix for solving the relevant elliptic equation.

- To bound material derivatives, we use estimates coming from the Euler-Reynolds equation itself and related commutator estimates to close the argument.

Considerations regarding the symmetries of the Euler equations, including scaling and Galilean transformations, also play an important role underlying the analysis. In particular, thanks to the ideas listed above, the bounds for the iteration depend only on relative velocities (i.e., derivatives of the velocity) but not on absolute velocities (i.e., the C^0 norm of the velocity).

As an interesting observation, it turns out that the total kinetic energy $\int \frac{|v|^2}{2}(t,x)dx$ for the solution obtained by the construction enjoys better regularity in time than what is proven for the solution itself (it is almost $C^{1/2}$ in t for the $C_{t,x}^{1/5-\delta}$ solutions we construct). In fact, in the conjectural "ideal case" scenario, the construction yields solutions whose energy functions are "almost automatically" in C^1 even though the velocity is only guaranteed to belong to $C_{t,x}^{1/3-\delta}$.

Using the same Main Lemma that is used to prove Theorem (0.1), we also prove the following theorem regarding the gluing of solutions.

Theorem 0.2. *For every smooth solution (v,p) to incompressible Euler on $(-2,2) \times \mathbb{T}^3$, there exists a Hölder continuous solution to Euler $(\bar{v},\bar{p})$ that coincides with (v,p) on $(-1,1) \times \mathbb{T}^3$ but is equal to a constant outside of $(-3/2, 3/2) \times \mathbb{T}^3$.*

Theorem (0.2) above was motivated by related discussions with P. Constantin, Y. Sinai, and T. Buckmaster.

1 The Euler-Reynolds System

We start by discussing some of the underlying philosophy behind the argument. A related discussion can be found in [DLSa].

Low Frequency Parts and Ensemble Averages of Euler Flows

Consider any solution (v_0, p_0) to the Euler equations on, say, $\mathbb{R}^n \times \mathbb{R}$, which we write in the form

$$\begin{cases} \partial_t v_0^l + \partial_j(v_0^j v_0^l) + \partial^l p_0 = 0 \\ \partial_j v_0^j = 0 \end{cases} \tag{7}$$

Here and in the rest of the paper, we use the Einstein summation convention, according to which we understand that there is a summation over the j index because it is repeated. Now imagine that v_0 could be very singular, such as one of the wild solutions that we will construct in the main body of the text, and consider how the "coarse scale" or "low frequency" part of v_0 might be allowed to behave.

One way to describe the coarse scale or low frequency part of the velocity field v_0 is to consider a mollification $v_\epsilon = \eta_\epsilon * v_0$. The mollifier η_ϵ in this argument could be a standard mollifier, so that the value of v_ϵ at each point is a weighted average of the velocity in an ϵ-neighborhood of the point, or the mollifier $\eta_\epsilon*$ could operate by projecting to spatial frequencies less than ϵ^{-1}. In either case, the result of the averaging process is to remove the fine scale oscillations in space at scales $|\Delta x| \leq \epsilon$, leaving only the coarse scale part of the velocity field as an average. By mollifying Equation (7), we see that the coarse scale velocity field $v = v_\epsilon$ and pressure $p = \eta_\epsilon * p_0$ satisfy the following system of equations, which are called the **Euler-Reynolds equations**.

$$\begin{cases} \partial_t v^l + \partial_j(v^j v^l) + \partial^l p = \partial_j R^{jl} \\ \qquad\qquad \partial_j v^j = 0 \end{cases} \tag{8}$$

The tensor $R^{jl} = v_\epsilon^j v_\epsilon^l - (v_0^j v_0^l)_\epsilon$ is a symmetric, non-positive $(2,0)$ tensor that arises from the failure of the nonlinearity to commute with the averaging. Since $v_0 \in L^2$, $R = R_\epsilon$ converges strongly to 0 in L^1 as $\epsilon \to 0$, and if v is continuous, R_ϵ converges to 0 in C^0. In general the rate at which R_ϵ converges to 0 and the function spaces in which this convergence occurs depend on the regularity of the solution v_0.

Actually, one can see that the average of **any** family of solutions to Euler will be a solution of the Euler-Reynolds equations. The most relevant type of averaging to convex integration arises during the operation of taking weak limits, which can be regarded as an averaging process as follows using the concept of a Young measure. For example, suppose that v_n is a sequence of solutions to the Euler equations uniformly bounded by $|v_n| \leq M$. Consider the sequence of measures $\mu_n(t, x, \tilde{v}) = dt dx \delta_{v_n}(\tilde{v})$ on $\mathbb{R}^n \times \mathbb{R} \times \mathbb{R}^n$ obtained by pushing forward the Lebesgue measure to the graph of v_n. Along some subsequence, which we also denote by v_n, the sequence of measures $\mu_n(t, x, \tilde{v})$ will obtain a weak limit which takes the form $dt dx d\mu_{t,x}(v)$. Measures that arise this way are called "Young measures" after being introduced by Young to describe the oscillatory behavior of minimizing sequences in the calculus of variations. The parameterized family of measures $\mu_{t,x}(v)$ are probability measures that record the local in space-time oscillations of the subsequence v_n. The center of mass $\bar{v}(t, x) = \int \tilde{v} d\mu_{t,x}(\tilde{v})$ of the Young measure is the weak limit of the subsequence v_n, and obeys the Euler-Reynolds system with Reynolds stress $R^{jl}(t, x) = \bar{v}^j \bar{v}^l - \int(\tilde{v}^j \tilde{v}^l) d\mu(\tilde{v})$. These latter statements follow from a general fact about Young measures: that is, for any continuous function $g(v)$

defined on $\{|v| \leq M\}$ the weak limit of $g(v_n)$ along the subsequence v_n exists and is given by the expected value $\int g(\tilde{v}) d\mu_{t,x}(v)$. While we do not use the theory of Young measures in this work, we find that their consideration is useful for visualizing and understanding the intuition behind the proofs in some of the previous literature and the one given here. We refer to [BHSM99] for more on the applications of Young measures.

Weak Limits of Euler Flows and the Hierarchy of Frequencies

In some sense, the convex integration procedure reverses the process in the above paragraph: one begins with a solution v_0 to the Euler-Reynolds system, and obtains a solution to the Euler equations by "reintroducing" the oscillations responsible for the forces that are exerted on v_0 "during" the weak limiting process. More precisely, the success of our procedure shows automatically that solutions to the Euler-Reynolds equations can be approximated in a weak topology by solutions to Euler. Since the method inherently proves that weak solutions to Euler may fail to be solutions, calculating what sort of velocity fields can be potentially realized as weak limits of solutions to Euler is important for finding a candidate space of subsolutions (or "approximate solutions") in which one can work while performing convex integration. However, it is only after the convex integration procedure is proven successful that we know the correct space has been found. A priori, one might be afraid that the class of solutions to the Euler-Reynolds system is too general to be entirely contained in the space of weak limits of Euler flows, since any incompressible flow v^l which conserves momentum can be regarded as a solution to the Euler-Reynolds equations after solving $\partial_j R^{jl} = \partial_t v^l + \partial_j(v^j v^l)$. In fact the conservation of momentum plays a subtle but important role in the construction, which we discuss further in Section (6.1). We now describe how convex integration is applied to Euler in more technical detail, but still from a general point of view.

The naive idea in constructing solutions through convex integration is to begin with the low frequency part of the solution (which is a smooth solution to the Euler-Reynolds equations), and then to "add in" the high frequency part of the solution one frequency shell at a time. This idea is implemented as follows. We begin with a smooth solution (v, p, R) of the Euler-Reynolds system (8), which we think of as the low frequency part of the solution we are trying to construct. We then add corrections to the velocity and pressure to obtain another (smooth) velocity field $v_1 = v + V$ and pressure $p_1 = p + P$, which together obey the Euler-Reynolds system but now with a new stress tensor R_1^{jl} that is much *smaller* than the previous stress R in an appropriate topology (for the present book, R will be measured in C^0). We iterate this procedure infinitely many times so that the stress tensor tends to 0 in the limit. In the process, the corrections to the velocity field become smaller in size in such a way that the sequence of velocity fields converges strongly in L^2 to some limit $\bar{v}$, which must be a weak solution to the Euler equations.

Once the construction is complete and the limiting solution $\bar{v}$ is known, we can look back on the process of the construction and see very explicitly the structure of the frequency levels of the solution. At each stage of the iteration, the correction V is chosen so that it oscillates rapidly compared to the velocity field v which came from the previous stage of the construction. As a result, the correction V we choose at each fixed stage of the iteration turns out to form a single, high frequency shell of the solution $\bar{v}$ that is constructed in the limit, whereas the velocity field v forms a low frequency projection of the solution $\bar{v}$. In this way, we have built the solution starting from the coarse scales and passing to fine scale oscillations, in a fashion similar to taking $\epsilon \to 0$ in the mollification procedure described above. With this intuition in mind, we expect the tensors R which appear to behave analogously to the family $R_\epsilon^{jl} = (v_\epsilon^j v_\epsilon^l) - (v_0^j v_0^l)_\epsilon$ that arises during the mollification process. In particular, as we seek solutions with C^α Hölder regularity, R should converge to 0 uniformly during the iteration at a rate that is consistent with the commutator estimate of [CET94] on R_ϵ.

Convex Integration and the h-principle for Weak Limits

The method of convex integration sharply contrasts other methods of solving PDE, as it usually leads to nonuniqueness results for the equations as well as to so-called "h-principle" results, which show that weak limits of solutions to the PDE may contain a large variety of non-solutions. In our context, this flexibility arises because the choice of the correction V at each stage permits many arbitrary choices, allowing one to obtain a huge family of solutions depending on these choices. In particular, our ability to show that weak limits of Euler flows may fail to be Euler flows comes ultimately from the freedom to choose an arbitrarily large frequency when constructing the correction V in the very first stage of the iteration (without affecting at all the boundedness or regularity of the solution obtained in the limit). Since every correction to follow will be of even more highly oscillatory nature, we find that the entire iteration results in a solution $\bar{v}$ that can be regarded as a perturbation of the velocity field v_0 that is taken in the first stage of the iteration. We conclude that the initial velocity field v_0 is a weak limit of solutions by choosing the initial frequency to be arbitrarily large.

These considerations explain not only why one must calculate the possible weak limits of Euler flows before attempting convex integration, but also why the results we achieve are connected to the h-principle. Namely, the method applied here always gives an approximation in a weak topology and cannot succeed unless the space of solutions is sufficiently abundant. This feature of the method is demonstrated in [Cho13], which contains a characterization of the H^{-1} closure of the space of $C_{t,x}^\alpha$ Euler flows for $\alpha < 1/10$ in 2 and 3 spatial dimensions.

Part II

General Considerations of the Scheme

We now highlight the main issues and the general philosophy underlying the proof of the Main Lemma. A part of this philosophy can also be found in [DLSa], since it also underlies the proof in [DLS13]. We use this section as an opportunity to introduce some heuristics which will be usefully formalized in the proof.

Suppose that (v, p, R) are a given solution of the Euler-Reynolds system.

$$\begin{cases} \partial_t v^l + \partial_j (v^j v^l) + \partial^l p = \partial_j R^{jl} \\ \qquad\qquad \partial_j v^j = 0 \end{cases} \tag{9}$$

For the purpose of the present discussion, let us imagine that v and p are smooth functions of size 5, and that R is a smooth, symmetric $(2,0)$ tensor field with absolute value smaller than one.

We introduce highly oscillatory corrections V and P to the velocity field and pressure such that $\operatorname{div} V = \partial_j V^j = 0$. The corrected velocity field $v_1 = v + V$ and pressure $p_1 = p + P$ satisfy the system

$$\partial_t v_1^l + \partial_j(v_1^j v_1^l) + \partial^l p_1 = \partial_t V^l + \partial_j(v^j V^l) + \partial_j(V^j v^l) + \partial_j(V^j V^l)$$
$$+ \partial^l P + \partial_j R^{jl}$$
$$\partial_j v_1^j = 0.$$

Our goal is to choose high frequency corrections V and P so that the forcing term in the equation can be represented as $\partial_j R_1^{jl}$ for a new Reynolds stress R_1^{jl} much smaller than R^{jl}.[2] Let us express the gradient $\partial^l P$ as a divergence $\partial_j(P\delta^{jl})$, where δ^{jl} is the (inverse) Euclidean inner product (or, if one prefers, the "Kronecker delta" or the "identity matrix"). We then collect terms as follows:

$$\partial_t v_1^l + \partial_j(v_1^j v_1^l) + \partial^l p_1 = \left[\partial_t V^l + \partial_j(v^j V^l) \right] + \left[\partial_j(V^j v^l) \right]$$
$$+ \partial_j \left[(V^j V^l) + P\delta^{jl} + R^{jl} \right]$$
$$\partial_j v_1^j = 0$$

We wish to express each of these terms in the form $\partial_j Q^{jl}$ with Q much smaller than $|R|$. Let us first consider the term

$$Q^{jl} = (V^j V^l) + P\delta^{jl} + R^{jl}. \tag{10}$$

---

[2]In general the word "smaller" here should refer to some norm that controls the $L^1_{t,x}$ norm, since we expect R to behave like the stress $R_\epsilon = (v_\epsilon^j v_\epsilon^l) - (v^j v^l)_\epsilon$ arising from mollifying a solution $v \in L^2_{t,x}$. In the present book, we will measure R in the norm $C^0_{t,x}$.

We cannot immediately make this term small because $V^j V^l$ is a rank one tensor, whereas $P\delta^{jl} + R^{jl}$ may be an arbitrary symmetric (2,0) tensor. We can, however, ensure that the low frequency part of Q^{jl} is small by choosing V so that the low frequency part of $V^j V^l$ cancels with the low frequency part of $P\delta^{jl} + R^{jl}$; to make sure this cancellation is possible, it is necessary to first choose P large enough so that $P\delta^{jl} + R^{jl}$ is negative definite, since the coarse scale part of $V^j V^l$ must be essentially positive definite. The fact that $V^j V^l$ can have a nontrivial low frequency part even though V itself has very high frequency demonstrates a lack of cancellation in the nonlinearity, which can ultimately be blamed for the fact that weak limits of Euler flows may fail to solve the Euler equation.[3] Note that the above choices of V and P imply that V (or, rather, the absolute value of V) has the size of a "square root" of R, and that P has the size of R.

$$|V| \approx |R|^{1/2}, \qquad |P| \approx |R|$$

Thus, after the construction is iterated k times with $R_{(k)} \to 0$ quickly enough, we can ensure that the velocities and pressures $v_{(k+1)} = v_{(k)} + V_{(k)}$, and $p_{(k+1)} = p_{(k)} + P_{(k)}$ constructed through the iteration will converge to solutions (v, p) of the Euler equations.

Although the low frequency part of the Q^{jl} term in (10) will be small, there will be many high frequency interaction terms in Q^{jl} which will not be small. So far, the only method available to handle these terms is to construct V out of stationary solutions to 3D-Euler called Beltrami flows, which also requires adding additional components to the pressure.[4]

Following [DLS13], we refer to the error term $\partial_t V^l + \partial_j(v^j V^l)$ as the "Transport term," since it can also be written as $(\partial_t + v^j \partial_j)V^l$. We want to find a small, symmetric tensor Q_T^{jl} solving the divergence equation

$$\partial_j Q_T^{jl} = (\partial_t + v^j \partial_j)V^l.$$

We also need to do the same for the "High-Low (frequency)" interaction term $\partial_j(V^j v^l)$, because $V^j v^l$ is of size "$|R|^{1/2}$," whereas we need Q to be much smaller than R. Achieving either of these goals requires us to find a small solution to the first order elliptic equation $\partial_j Q^{jl} = U^l$, with high frequency data U^l.

In general, we expect that if U^l is a smooth vector field with amplitude of "size 1" and with "frequency λ," solving the elliptic equation should allow us to achieve a solution of "size $1/\lambda$." One way to give rigorous evidence to this heuristic expectation is to use the Fourier transform to solve the

---

[3] Regarding the analogous aspect of the construction in [Shn97], Shnirelman suggested that the above idea could be thought of as emulating the frequency cascades predicted in the theory of turbulence.

[4] For the isometric embedding problem, there is no known method to handle interference terms, and this difficulty actually limits the regularity of the solutions which can currently be obtained through convex integration. See [CDLS12b].

equation (see Section (6)). More evidence can be drawn by making an analogy with the ODE $\frac{dQ}{dx} = e^{i\lambda x}u(x)$, where repeated integration by parts yields the estimate,

$$\frac{dQ}{dx} = e^{i\lambda x}u(x) \tag{11}$$

$$\Rightarrow \|Q\| \lesssim \frac{\|u\|}{\lambda} + \frac{\|\nabla u\|}{\lambda^2} + \ldots + \frac{\|\nabla^D u\|}{\lambda^D} + \frac{\|\nabla^{D+1} u\|}{\lambda^D}. \tag{12}$$

Thus, with the appropriate "integration by parts" type argument, we expect to gain a smallness factor of $\|Q\| \lesssim \frac{\|u\|}{\lambda}$ by solving the divergence equation $\partial_j Q^{jl} = e^{i\lambda \xi \cdot x} u^l$, as long as the oscillations of u are slower than λ (i.e., u cannot be of the form $e^{-i\lambda \xi \cdot x} v^l$ for some slowly varying v).

The Transport term presents a more serious problem than does the High-Low term $\partial_j(V^j v^l) = V^j \partial_j v^l$, because the Transport term necessarily involves differentiating the oscillatory correction V and then solving the equation

$$\partial_j Q_T^{jl} = (\partial_t + v^j \partial_j) V^l. \tag{13}$$

Now, v has size 1, and if V has the expected size $|R|^{1/2}$ and frequency λ, its derivative $v^j \partial_j V$ on the right hand side will have size at least $\lambda |R|^{1/2}$ and will also have frequency λ. On the other hand, solving the elliptic equation only gains one power of λ^{-1}, leaving no hope to find a solution with $|Q| < |R|$ when $|R|$ is small. The equations therefore impose a requirement that V be essentially transported by the coarse scale velocity field v.

The philosophy described above was executed in [DLS13] to construct continuous, weak solutions of Euler which do not conserve energy (with the superficial difference that the stress R^{jl} was required to have trace $R^{jl}\delta_{jl} = 0$). However, while the paper [DLS13] was the first to execute this philosophy by representing the forcing term as a divergence $\partial_j R^{jl}$, many of the above considerations also underlie the preceding constructions of badly behaved weak solutions to Euler, including the earlier examples of [Sch93], [Shn97] and [DLS09].

2   Structure of the Book

While the general considerations above pertain to both the present book and to the construction in [DLS13], the manner in which the philosophy is executed in the present book is different. Our strategy for the exposition is as follows.

- In Section (3), we identify the error terms which need to be controlled.

- In Part (III) we explain some notation of the book and write down a basic construction of the correction. In this part, we do not explain how the various parameters involved should be chosen to optimize the construction. Rather, our goal is to give enough detail so that it is clear how the scheme can be used to construct weak solutions which are continuous in space and time.

- In Part (IV) we explain how to iterate the construction of Part (III) to obtain continuous solutions to the Euler equations. We then explain the concept of frequency and energy levels, and the Main Lemma which is iterated to give the full proof of the Theorems (0.1) and (0.2). Through this discussion, we explain some additional difficulties which present themselves as one approaches the optimal regularity, and how these difficulties can be overcome by isolating the material derivative $\partial_t + v \cdot \nabla$ as a special derivative.

- In Parts (V)–(VII), we verify all the estimates needed for the proof of the Main Lemma.

3 Basic Technical Outline

We now give a more technical outline of the construction.

Let (v^l, p, R^{jl}) be a solution to the Euler-Reynolds system, and consider a correction $v_1 = v + V$, $p_1 = p + P$.

The correction V is a divergence free vector field which oscillates rapidly compared to v, and can be written as a sum of divergence free vector fields $V = \sum_I V_I$, which oscillate in various different directions, and which are supported in several different regions of the space-time $\mathbb{T}^3 \times \mathbb{R}$.

In our construction, there will always be a bounded number of waves V_I (at most 192) which are nonzero at any given time t.

Each individual wave V_I composing V is a complex-valued, divergence free vector field that oscillates rapidly in only one direction. We introduce several ways in which we will represent each V_I:

$$V_I = e^{i\lambda \xi_I} \tilde{v}_I$$

$$= \frac{1}{\lambda} \nabla \times (e^{i\lambda \xi_I} w_I)$$

$$= e^{i\lambda \xi_I} (v_I + \frac{\nabla \times w_I}{\lambda})$$

$$v_I = (i\nabla \xi_I) \times w_I$$

Here, $\lambda \in \mathbb{R}$ is a large number, independent of (t, x), which will be chosen in the argument. The phase functions $\xi_I(t, x)$ will be real-valued functions of (t, x) whose gradients indicate the direction of oscillation of the components V_I. Unlike the argument of [DLS13], we will not require the functions $\xi_I(t, x)$ to be linear (that is, to have constant gradients).

Because we use nonlinear phase functions, we must modify the "nonstationary phase" type arguments used in [DLS13] to gain cancellation when solving $\partial_j Q^{jl} = e^{i\lambda\xi(x)}u^l$. We therefore must ensure, for one thing, that the gradients $\partial^l\xi_I$ of the phase functions ξ_I remain bounded away from 0. If one is willing to restrict to $\lambda \in \mathbb{Z}$ (as is done in [DLS13]), one could obtain globally defined functions ξ_I taking values in $\mathbb{R}/(2\pi\mathbb{Z})$ which satisfy this lower bound on their gradients. Instead, we choose to define ξ_I on only part of the space $\mathbb{T}^3$ at any given time, and the coefficient $\tilde{v}_I^l$ will be a vector field that is compactly supported in that region, and also compactly supported in time. In particular, we will not be using the globally defined frequencies on the torus.

The $\tilde{v}_I = \tilde{v}_I^l$, and $w_I = w_I^l$ here are complex-valued vector fields. Then $v_I = (i\nabla\xi_I) \times w_I$ is a vector field which will be chosen later, but which necessarily takes values which point into the level sets of ξ_I. Since λ will be a large parameter, $\tilde{v}_I \cdot \nabla\xi_I = \tilde{v}_I^l\partial_l\xi_I = \frac{\nabla\times w_I}{\lambda} \cdot (\nabla\xi_I)$ will be small. The character of these building blocks reflects the fundamental principle that any high frequency, divergence free plane wave must point perpendicular to its direction of oscillation. This geometric constraint causes the vector field to generate a shearing type of flow. This geometric principle for divergence free plane waves can be formalized on $\mathbb{R}^n$ by considering the Fourier transform, and still holds approximately in our setting where we consider nonlinear phases. In our argument, we will first construct the vector field v_I, and then solve the linear equation $v_I = (i\nabla\xi_I) \times w_I$ in order to obtain w_I.

We remark that, in order to ensure that V is real-valued, each index I will have a conjugate index $\bar{I}$ such that $V_{\bar{I}} = \bar{V}_I$. Moreover, $\xi_{\bar{I}} = -\xi_I$, and $w_{\bar{I}} = \bar{w}_I$ so that $v_{\bar{I}} = \bar{v}_I$ as well.

Let us introduce the notation $(vw)^{jl}$ to represent the symmetric product of two vectors v^l and w^l. That is, $(vv)^{jl} = v^jv^l$ is the square $v \otimes v$ of v, and from the polarization identity $(vw)^{jl} = \frac{1}{2}(v^jw^l + w^jv^l)$ in general. The symmetric product so defined is commutative.

Using the representation of V as a sum of individual waves with distinct direction and location, we see that the corrected velocity field and pressure v_1, p_1 satisfy the system.

$$\partial_t v_1^l + \partial_j(v_1^j v_1^l) + \partial^l p_1 = [\partial_t V^l + \partial_j(v^j V^l)] + [\partial_j(V^j v^l)]$$

$$+ \sum_{J \neq \bar{I}} \partial_j(V_I V_J)^{jl} + \partial_j\left[\sum_I (V_I^j \bar{V}_I^l) + P\delta^{jl} + R^{jl}\right] \quad (14)$$

$$\partial_j v_1^j = 0 \quad (15)$$

We have separated out the interaction terms between non-colliding frequencies in the hope that these interference terms can be shown to be negligible. Indeed, if one measures errors in a weak topology, these interaction terms will be small because the products $(V_I V_J)^{jl}$ can easily be made high frequency for $J \neq \bar{I}$; furthermore, the term $\sum_I (V_I^j \bar{V}_I^l) + P\delta^{jl} + R^{jl}$ will become small in a strong topology as long as the amplitudes v_I are chosen

15

appropriately.[5] However, we cannot use a weak topology to measure the size of the remaining stress.[6]

Ultimately, we will only be able to handle the interaction terms after imposing a "Beltrami flow" condition on the structure of the V_I. This condition will allow us to show that after adding appropriate terms to the pressure, the interference terms $(V_I V_J)^{jl} - P_{IJ}\delta^{jl}$ are small in C^0 modulo solutions of $\partial_j Q^{jl} = 0$.

The pressure will therefore be of the form

$$P = P_0 + \sum_{J \neq \bar{I}} P_{I,J}$$

where P_0 appears in the equation

$$\sum_I V_I^j \bar{V}_I^l + P_0 \delta^{jl} + R^{jl} \approx 0. \tag{16}$$

We note here that, by choosing P_0 appropriately, we will also be able to prescribe the energy increment

$$\int (|v + V|^2 - |v|^2) dx \approx e(t)$$

of the correction with great accuracy.

There are two reasons for the $\approx$ symbol in (16). One reason is that we cannot control the pointwise values of

$$V_I^j \bar{V}_I^l = \tilde{v}_I^j \bar{\tilde{v}}_I^l$$

exactly; rather we are only able to determine these products approximately in the sense that

$$V_I^j \bar{V}_I^l \approx v_I^j \bar{v}_I^l$$

and we have freedom to prescribe v_I up to some constraints such as the fact that $v_I \in \langle \nabla \xi_I \rangle^\perp$.

The other reason for the approximation symbol $\approx$ in (16) is that we will also define a suitable mollification R_ϵ of R before solving the equation

$$\sum_I v_I^j \bar{v}_I^l + P_0 \delta^{jl} + R_\epsilon^{jl} = 0 \tag{17}$$

---

[5]One can interpret this construction as requiring that $V^l = V_{(\lambda)}^l$ generates an appropriate Young measure as $\lambda \to \infty$. The fact that this equation can be solved for arbitrary R, implying in a sense that high frequencies may emulate arbitrary forces on the lower frequency part of the solution, demonstrates a lack of cancellation characteristic of the nonlinearity in the Euler equations.

[6]Recall from Section 1 that we expect $R_\epsilon^{jl} \approx (v^j v^l)_\epsilon - v_\epsilon^j v_\epsilon^l$ to vanish uniformly if v is continuous, and must at least converge to 0 in L^1 if v belongs to L^2.

pointwise. In this way, the amplitudes v_I will only carry coarse scale information regarding R.

Making this mollification introduces another error term of the form

$$R - R_\epsilon$$

which will be carried as part of the new stress R_1. For similar reasons, we will also introduce an appropriate mollification v_ϵ of v at the beginning of the argument, giving rise to another term

$$(v^j - v_\epsilon^j)V^l + V^j(v^l - v_\epsilon^l)$$

in the new stress. While these mollifications are very important for the construction of Hölder continuous solutions, for the present discussion we will ignore them, and return to them later.

To summarize, we have seen five main error terms, each of which must be expressed in the form $\partial_j Q^{jl}$ for some $Q^{jl}(t, x)$ taking values in the space of symmetric, $(2, 0)$ tensors $\mathcal{S}$, and each Q^{jl} is required to be smaller than R^{jl} to ensure we have made visible progress towards a solution of the Euler equations. The terms we must deal with are named:

- **The Transport Term**

$$\partial_j Q_T^{jl} = \partial_t V^l + \partial_j(v_\epsilon^j V^l)$$

- **The High-Low Interaction Term**

$$\partial_j Q_L^{jl} = \partial_j(V^j v^l) = V^j \partial_j v_\epsilon^l$$

- **The High-High Interference Terms**

$$\partial_j Q_H^{jl} = \sum_{J \neq \bar{I}} \partial_j \left[(V_I V_J)^{jl} + P_{I,J} \delta^{jl} \right]$$

- **The Stress Term**

$$Q_S^{jl} = \left[\sum_I (V_I^j \bar{V}_I^l) + P_0 \delta^{jl} + R_\epsilon^{jl} \right]$$

- **The Mollification Terms**

$$Q_M^{jl} = (v^j - v_\epsilon^j)V^l + V^j(v^l - v_\epsilon^l) + (R^{jl} - R_\epsilon^{jl})$$

In [DLS13], the High-High Interference terms and Stress term both correspond to parts of what is called the "oscillation part" of the error.

Each of these terms has its own set of difficulties, and these difficulties are coupled together as the Transport term and Stress term both impose constraints on the correction V. The one difficulty that is common to the first three of the terms is that they require one to find small solutions to the underdetermined, first-order, elliptic equation

$$\partial_j Q^{jl} = e^{i\lambda\xi(t,x)} u^l(t,x).$$

For example, the High-Low Interaction term (which is the simplest of these three error terms) can be expanded in the form

$$V^j \partial_j v_\epsilon^l = \sum_I e^{i\lambda\xi_I} \tilde{v}_I^j \partial_j v_\epsilon^l.$$

As we previously remarked, either by considering the problem in frequency space or by drawing an analogy with the ODE $\frac{dQ}{dx} = e^{i\lambda\xi(x)} u(x)$, one expects to gain a smallness factor of $\frac{1}{\lambda}$ for the solution Q. This gain is achieved by an "oscillatory estimate," whose proof we will describe shortly.

This gain of λ^{-1} alone is not sufficient to handle every term in the argument; both the Transport term and the High-High Interference terms require one to differentiate the oscillatory correction V, and thereby produce oscillatory data whose main terms are of the form $e^{i\lambda\xi}\lambda u^l$, meaning their amplitudes are too large for the oscillatory estimate to obtain a small solution. For example, the right-hand side of the Transport term can be written as

$$\begin{aligned}
\partial_t V^l + \partial_j(v_\epsilon^j V^l) &= (\partial_t + v_\epsilon^j \partial_j) V^l \\
&= \sum_I e^{i\lambda\xi_I} \left(i\lambda(\partial_t + v_\epsilon^j \partial_j)\xi_I \tilde{v}_I^l + (\partial_t + v_\epsilon^j \partial_j)\tilde{v}_I^l \right).
\end{aligned}$$

For the Transport term, we deal with this difficulty by allowing the phase functions ξ_I to obey a transport equation; in particular, the phase functions will not be linear functions, in contrast to the argument in [DLS13].

We will also make sure the lifespan of each V_I is a sufficiently small time interval, so that phase functions ξ_I remain nonstationary and suitably regular for applying the main, oscillatory estimate. The length of this time interval will depend on bounds for the derivatives of v, including the L^∞ norm of ∇v.

Part III

Basic Construction of the Correction

4 Notation

We employ the Einstein summation convention, according to which there is an implied summation when a pair of indices is repeated (e.g., $\partial_j v^j$ is the divergence of a vector field, $(v \cdot \nabla)f = v^j \partial_j f$, etc.). We employ the conventions of abstract index notation, so that upper indices and lower indices distinguish contravariant and covariant tensors.

We define the space $\mathcal{S} = \mathrm{Sym}^2(\mathbb{R}^3)$ to be the six dimensional space of symmetric, $(2,0)$ tensors on $\mathbb{R}^3$. That is, the vectors in $\mathcal{S}$ are symmetric bilinear maps $G^{jl} : (\mathbb{R}^3)^* \times (\mathbb{R}^3)^* \to \mathbb{R}$ on the dual of $\mathbb{R}^3$, whose action on a pair of covectors $u, v \in (\mathbb{R}^3)^*$ can be written as $G(u, v) = G^{jl} u_j v_l = G^{lj} u_j v_l = G(v, u)$. We also define $\mathcal{S}_+$ to be the cone of positive definite, symmetric, $(2,0)$ tensors.

We will use the notation

$$(vu)^{jl} = \frac{1}{2}(v^j u^l + v^l u^j)$$

to refer to the symmetric product of two vectors v and u.

At many points, we will write inequalities of the form

$$X \trianglelefteq Y.$$

The symbol $\trianglelefteq$ expresses that the above inequality is a goal, and that at some point later on in the proof we will have to show that the inequality $X \leq Y$ does in fact hold.

The following notation concerning multi-indices will later be helpful for expressing higher order derivatives of a composition (see Section (18.4)).

Definition 4.1. A K-tuple of multi-indices $(a^1, a^2, \ldots, a^K)$ is said to form an ordered K-partition of a multi-index $a = (a_1, a_2, \ldots, a_N)$ if there is a partition $\{1, \ldots, N\} = \pi_1 \cup \cdots \cup \pi_K$ with the following properties:

- The subsets $\pi_j \subseteq \{1, \ldots, N\}$, $j = 1, \ldots, K$ are pairwise disjoint

- Each multi-index a^j has the form

$$(a_1^j, a_2^j, \ldots, a_{N_j}^j) = (a_{\pi_j(1)}, a_{\pi_j(2)}, \ldots, a_{\pi_j(N_j)})$$

  where $\pi_j = \{\pi_j(1), \ldots, \pi_j(N_j)\}$ is written in increasing order, and

- The subsets are ordered by their largest elements

$$\pi_1(N_1) < \pi_2(N_2) < \ldots < \pi_K(N_K).$$

19

5 A Main Lemma for Continuous Solutions

In order to have a concrete goal for the construction, we state a lemma that implies the existence of continuous solutions.

Lemma 5.1 (Main Lemma for Continuous Solutions). *There exist constants K and C such that the following holds.*

Let $\epsilon > 0$, and suppose that (v, p, R) are uniformly continuous solutions to the Euler-Reynolds equations on $\mathbb{R} \times \mathbb{T}^3$, with v uniformly bounded[7] and

$$\text{supp } R \subseteq I \times \mathbb{T}^3$$

for some time interval, and

$$\|R\|_{C^0_{t,x}} \leq e_R.$$

Let

$$e(t) : \mathbb{R} \to \mathbb{R}_{\geq 0}$$

be any function satisfying the lower bound

$$e(t) \geq K e_R \tag{18}$$

on a neighborhood of I such that

$$\frac{d}{dt} e^{1/2}(t) \in C^0(\mathbb{R}) \tag{19}$$

is continuous and uniformly bounded, and such that $e(t)$ is bounded by

$$e(t) \leq 1000 K e_R. \tag{20}$$

Then there exists a uniformly continuous solution (v_1, p_1, R_1) to the Euler-Reynolds equations of the form

$$v_1 = v + V \tag{21}$$
$$p_1 = p + P \tag{22}$$

such that the supports of the corrections and the new stress

$$\text{supp } V \cup \text{ supp } P \cup \text{ supp } R_1 \subseteq \text{ supp } e \times \mathbb{T}^3 \tag{23}$$

and so that V and P obey the bounds

$$\|V\|_{C^0} \leq C e_R^{1/2} \tag{24}$$
$$\|P\|_{C^0} \leq C e_R \tag{25}$$

---

[7]The assumption of uniform boundedness here may be somewhat unnatural and actually does not enter into the construction of Hölder continuous solutions, where all the bounds in the construction depend only on relative velocities.

and

$$\left\| \int |v_1|^2(t,x)dx - \int (|v|^2 + e(t))dx \right\|_{C_t^0} \leq \epsilon \tag{26}$$

$$\|R_1\|_{C^0} \leq \epsilon. \tag{27}$$

It is not difficult to check that Lemma 5.1 implies the following theorem:

Theorem 5.1. *There exist continuous solutions (v,p) to the Euler equations that are nontrivial and have compact support in time.*

Proof of Theorem (5.1). To establish Theorem (5.1), one repeatedly applies Lemma (5.1) to produce a sequence of solutions to the Euler-Reynolds equations $(v_{(k)}, p_{(k)}, R_{(k)})$ for which the stress is bounded by

$$\|R_{(k)}\|_{C^0} \leq e_{R,(k)} \tag{28}$$

with $e_{R,(k)} > 0$ a decreasing sequence of positive numbers chosen to tend to 0 rapidly. The corrections to the velocity $v_{(k+1)} = v_{(k)} + V_{(k)}$ and to the pressure $p_{(k+1)} = p_{(k)} + P_{(k)}$ can be summed in C^0

$$\sum_k \|V_{(k)}\|_{C^0} \leq C \sum_k e_{R,(k)}^{1/2} < \infty \tag{29}$$

$$\sum_k \|P_{(k)}\|_{C^0} \leq C \sum_k e_{R,(k)} < \infty \tag{30}$$

by (24) and (25), which implies the uniform convergence of

$$v = \lim_k v_{(k)}$$

$$p = \lim_k p_{(k)}$$

to uniformly continuous solutions (v,p) of the Euler equations.

To make sure the solutions constructed in this way are nontrivial and compactly supported, we apply the lemma with $e(t)$ chosen to be any sequence of non-negative functions $e_{(k)}(t)$ that satisfy (18), (19) and (20) for $e_R = e_{R,(k)}$ and whose supports are all contained in some finite time interval. We can also arrange that these functions are strictly positive at time 0 so that

$$e_{(k)}(0) > 0$$

for all k.

If we choose $\epsilon_{(k)}$ small enough at each stage, the inequality (26) implies that the energy at $t = 0$

$$\int |v_{(k+1)}|^2(0,x)dx \geq \int (|v_{(k)}|^2(0,x) + e_{(k)}(0))dx - \epsilon_{(k)} \tag{31}$$

$$> \int |v_{(k)}|^2(0,x)\, dx \tag{32}$$

increases with each stage. Then, by the dominated convergence theorem,

$$\int |v|^2(0,x)dx = \lim_k \int |v_{(k)}|^2(0,x)\,dx \tag{33}$$

$$\geq \int |v_{(1)}|^2(0,x)dx > 0 \tag{34}$$

which ensures that the solution $v = \lim_k v_{(k)}$ is nonzero at $t = 0$. $\qquad\square$

Let us now proceed with the construction. During the construction, we will not be completely specific about certain technical points (such as defining precisely how to mollify v and R) because these aspects must be changed or handled more delicately in order to construct the Hölder continuous solutions of Theorem (0.1). Instead we will keep the construction stated in general terms so that we can refer to the same proof for both the continuous and Hölder continuous cases.

6   The Divergence Equation

A key ingredient in the proof of Lemma (5.1) is to find special solutions to the divergence equation

$$\partial_j Q^{jl} = e^{i\lambda\xi(x)}u^l \tag{35}$$

where ξ and u^l are respectively a smooth function and a smooth vector field on $\mathbb{T}^3$, and Q^{jl} is an unknown, symmetric $(2,0)$ tensor. In our applications, u^l will always be supported in a single coordinate chart, and ξ will only be defined in a neighborhood of the support of u^l.

We will need to take advantage of the highly oscillatory nature of the data on the right-hand side of (35), as we expect to gain a smallness factor of $\frac{1}{\lambda}$ for the solution Q. The estimate we obtain (Proposition 6.1) is analogous to the one-dimensional estimate

$$\|Q\|_{C^0(\mathbb{R}/\mathbb{Z})} \leq C\left(\frac{\|u\|_{C^0(\mathbb{R}/\mathbb{Z})}}{\lambda} + \frac{\|\nabla u\|_{L^1(\mathbb{R}/\mathbb{Z})}}{\lambda}\right)$$

which one can prove for periodic solutions to the equation $\frac{dQ}{dx} = e^{i\lambda\xi(x)}u(x)$ using integration by parts. To obtain this cancellation, a key point is that the phase function must remain nonstationary, and more precisely one requires bounds on $\||\xi'(x)|^{-1}\|_{C^0}$ and $\|\xi''(x)\|_{L^1}$ to obtain the oscillatory estimate in the one-dimensional model. Likewise, Proposition 6.1 will require similar assumptions on the phase functions.

6.1   A Remark about Momentum Conservation

A first remark regards why one cannot solve the equation

$$\partial_j Q^{jl} = U^l$$

for arbitrary data on the right-hand side. Namely, the right-hand side must have integral 0 in order to be a divergence. This condition is also sufficient, but let us briefly discuss how this condition relates to the conservation of momentum.

From a physical point of view, the right-hand side of the Euler equation is a force, and the condition

$$\int U^l dx = 0$$

reflects Newton's law that every action must have an equal and opposite reaction. This axiom, in turn, implies the conservation of momentum in classical mechanics.

When translated into the general Hilbert space framework for solving elliptic equations, the condition required on the data $U^l = e^{i\xi(x)}u^l$ is that U^l be orthogonal to the kernel of the operator adjoint to the divergence operator $\mathcal{P}(Q) = \partial_j Q^{jl}$. Here the operator adjoint to $\mathcal{P}$ is the operator $\mathcal{P}^*(v) = -\frac{1}{2}(\partial^j v^l + \partial^l v^j)$, which is essentially the symmetric part of the derivative. The tensor $\partial^j v^l + \partial^l v^j$ has the familiar geometric interpretation as a deformation tensor of the vector field v — that is, it expresses the Lie derivative $\mathcal{L}_v \delta^{jl}$ of the inner product δ^{jl} when pulled back along the flow of v. Those vector fields whose deformation tensor $\partial^j v^l + \partial^l v^j$ vanishes are called Killing vector fields, and they consist of those vector fields whose flows generate isometries of the space. On the torus, these Killing vector fields consist of the translation vector fields; that is, vector fields whose component functions v^l are constant on $\mathbb{T}^3$. The integral 0 condition is the condition that v be orthogonal to translations.

In view of Noether's theorem, the constant vector fields which act as Galilean symmetries of the Euler equation are responsible for the conservation of momentum. In this case, conservation of momentum for solutions to Euler is proven directly by integrating any component of the equation

$$\partial_t v^l + \partial_j(v^j v^l) + \partial^l p = 0$$

in space.

As the proof reveals, **all** solutions to the Euler equations, even those that only belong to L^2, conserve momentum. In fact, by the same proof for the equations

$$\partial_t v^l + \partial_j(v^j v^l) + \partial^l p = \partial_j R^{jl}$$

all solutions to the Euler-Reynolds equations also conserve momentum. Conversely, any incompressible flow that conserves momentum can be realized as a solution to the Euler-Reynolds equations by solving the divergence equation $\partial_j R^{jl} = \partial_t v^l + \partial_j(v^j v^l)$. The orthogonality condition implies that only those forces U^l that do not inject any momentum into the system can be realized as the divergence of a stress $\partial_j Q^{jl}$, thereby guaranteeing the conservation of momentum for the solutions we construct.

23

This condition will not present any difficulty for the present argument, since all the terms U^l which arise when we introduce the correction $v_1 = v + V$ satisfy the orthogonality condition. For example, the data for High-Low Interaction term

$$\partial_j Q_L^{jl} = \partial_j(V^j v_\epsilon^l)$$

has integral 0 since it is a divergence, and the data for the Transport term

$$\partial_j Q_T^{jl} = \partial_t V + \partial_j(v_\epsilon^j V^l)$$

has integral 0 because

$$\partial_t V = \nabla \times \partial_t W$$

and the curl of any vector field on $\mathbb{T}^3$ has integral 0.

6.2   The Parametrix

Having discussed the compatibility condition for solving the equation, let us now prove the main oscillatory estimate on the solution. To convey the idea, we first prove a simple (C^0) version. To be consistent with the notation of the book, we keep a scalar λ in the statement of the theorem, although λ can without loss of generality be absorbed into the phase function if one prefers.

Proposition 6.1. *Let U^l be a smooth vector field on $\mathbb{T}^3$ with the property that $\int_{\mathbb{T}^3} U^l(x)dx = 0$, and suppose that U^l can be represented as $e^{i\lambda\xi(x)}u^l$ for some smooth vector field u^l and some smooth, real-valued phase function ξ defined in a neighborhood of the support of u^l whose gradient does not vanish at any point. Then there exists a symmetric, $(2,0)$ tensor field Q^{jl} on $\mathbb{T}^3$ solving the equation*

$$\partial_j Q^{jl} = U^l = e^{i\lambda\xi(x)}u^l,$$

which depends linearly on U^l and for any $p > 3$, $0 \le s \le 1$ satisfies the estimate

$$||Q^{jl}||_{C^0(\mathbb{T}^3)} \le C\left(\frac{||\ |\nabla\xi|^{-1}\ ||_{L^\infty}(1+||\ |\nabla\xi|^{-1}\ ||_{L^\infty}||D^2\xi||_{L^p})}{\lambda}\right)^s ||u||_{W^{s,p}}.$$

The implied constant depends only on s and p.

The $s = 0$ case derives from the following, standard elliptic estimate, which is used as a lemma in the proof.

Lemma 6.1. *If f^l be a smooth vector field on $\mathbb{T}^3$ such that $\int f^l dx = 0$, there exists a symmetric, $(2,0)$ tensor field Q^{jl} depending linearly on f^l which solves the equation*

$$\partial_j Q^{jl} = f^l$$

24

and satisfies the bound

$$||Q^{jl}||_{W^{1,p}(\mathbb{T}^3)} \leq C||f||_{L^p(\mathbb{T}^3)}$$

so that, in particular when $p > 3$

$$||Q^{jl}||_{C^0(\mathbb{T}^3)} \leq C||f||_{L^p(\mathbb{T}^3)}.$$

The theorem itself provides an interpolation between the C^0 estimate in the lemma, and the $s = 1$ bound. In fact, the $s = 1$ case of the following lemma is sufficient for the construction of continuous weak solutions, but we prove the proposition for fractional regularity in order to demonstrate the robustness of the method, and because the proof illustrates an important theme of the paper.

The idea of the proof, which generalizes the approach taken in [DLS13] to nonlinear phase functions, is as follows. When using the Fourier transform to solve this equation, the key observation to be made is that whenever u^l is constant, and $\xi(x)$ is linear (so that the derivatives $\partial_j \xi$ are constant), an exact solution can be obtained by the formula

$$Q^{jl} = \frac{1}{\lambda}\left(e^{i\lambda\xi(x)}q^{jl}\right)$$

for any constant tensor q^{jl} solving the linear equation

$$i\partial_j \xi q^{jl} = u^l$$

pointwise. In fact, one can arrange that the solution q^{jl} is given explicitly by a map

$$q^{jl} = q^{jl}(\nabla\xi)[u] \tag{36}$$

that is homogeneous of degree -1 in $\nabla\xi$ and is linear in u.

If $\partial_j \xi$ and u^l are not constant, but are still smooth, one can still obtain an approximate solution of this form, and eliminating the error requires one to solve $\partial_j Q^{jl} = f^l$, where the data f^l is small in L^p, allowing us to apply Lemma (6.1).

We now proceed with the formal proof:

Proof. The preceding discussion suggests that we begin with an approximate solution $e^{i\lambda\xi(x)}q^{jl}$, where q^{jl} is chosen to solve the linear equation $i\partial_j \xi q^{jl} = u^l$ pointwise, but doing so is only a good idea when u^l is smooth. Therefore, we find an approximate solution to $\partial_j Q^{jl} = e^{i\lambda\xi}u^l_\epsilon$, where $u^l_\epsilon = \eta_\epsilon * u^l$ is a standard mollification of u^l. As is typical of stationary phase arguments, the parameter ϵ will be optimized later to ensure that the oscillations of the phase function are rapid compared to the coarser scale variations of u^l_ϵ; a mollification in a similar spirit will be used at many other instances in the main body of the book. In any case, we do not expect

25

to observe cancellation here unless the "frequency of u" is less than that of $\lambda\xi$; otherwise u^l could be of the form $e^{-i\lambda\xi(x)}v^l$.

To solve the linear equation $i\partial_j\xi q_\epsilon^{jl} = u_\epsilon^l$, we first decompose $u_\epsilon^l = u_\perp^l + \frac{(u_\epsilon \cdot \nabla\xi)}{|\nabla\xi|^2}\partial^l\xi = u_\perp^l + u_\parallel^l$, where $(u_\perp) \cdot \nabla\xi = 0$, and $u_\parallel$ points in the direction of $\nabla\xi$. We then set q_ϵ^{jl} to be the sum

$$q_\epsilon^{jl} = i^{-1}(q_\perp^{jl} + q_\parallel^{jl}) \tag{37}$$

$$= q^{jl}(\nabla\xi)[u_\epsilon] \tag{38}$$

where

$$q_\perp^{jl} = \frac{1}{|\nabla\xi|^2}(\partial^j\xi u_\perp^l + \partial^l\xi u_\perp^j)$$

so that $\partial_j\xi q_\perp^{jl} = u_\perp^l$, and

$$q_\parallel^{jl} = \frac{(u_\epsilon \cdot \nabla\xi)}{|\nabla\xi|^2}\delta^{jl}$$

so that $\partial_j\xi q_\parallel^{jl} = u_\parallel^l$.

We now seek a solution of the form $Q^{jl} = \frac{1}{\lambda}e^{i\lambda\xi}q_\epsilon^{jl} + Q_1^{jl}$ where Q_1^{jl} must satisfy the equation:

$$\partial_j Q_1^{jl} = f^l$$

with the data $f^l = -e^{i\lambda\xi(x)}((u^l - u_\epsilon^l) + \frac{1}{\lambda}\partial_j q_\epsilon^{jl})$.

Since the error term on the right-hand side is still of the form $e^{i\lambda\xi}\bar{u}^l$, it is possible to iterate the preceding method in order to produce a higher order approximate solution at the price of taking more derivatives of ξ and u^l. Later in the book, we will use a higher order parametrix.

For now, we will simply observe that the remaining data can already be regarded as bounded in L^p. The right-hand side also has integral zero because it is of the form $U^l + \partial_j\tilde{Q}^{jl}$, and the original data U^l was assumed to have integral 0. Therefore, by Lemma (6.1), there exists a smooth solution Q_1^{jl} on $\mathbb{T}^3$ obeying the estimate

$$\|Q_1^{jl}\|_{C^0(\mathbb{T}^3)} \lesssim \|e^{i\lambda\xi(x)}((u^l - u_\epsilon^l) + \frac{1}{\lambda}\partial_j q_\epsilon^{jl})\|_{L^p(\mathbb{T}^3)}$$

$$\lesssim \|(u - u_\epsilon)\|_{L^p} + \frac{1}{\lambda}\|\partial_j q_\epsilon^{jl}\|_{L^p}$$

$$\lesssim \epsilon^s\|u\|_{\dot{W}^{s,p}} + \frac{1}{\lambda}\|\partial_j q_\epsilon^{jl}\|_{L^p}.$$

We will not be using any special structure involving the divergence, so one might as well take the entire norm $\|\nabla q_\epsilon\|_{L^p}$. Let us estimate the size of $\partial_j q_\epsilon^{jl}$ as follows. By differentiating the system of equations

$$|\nabla\xi|^2 q_\epsilon^{jl} = (\partial^j\xi u_\perp^l + \partial^l\xi u_\perp^j) + (u_\epsilon\cdot\nabla\xi)\delta^{jl}$$
$$|\nabla\xi|^2 u_\epsilon^l = |\nabla\xi|^2 u_\perp^l + (u_\epsilon\cdot\nabla\xi)\partial^l\xi,$$

one computes that $\partial_j q_\epsilon^{jl}$ as a linear combination of two different types of terms:

- Terms with L^p norm bounded by $||\,|\nabla\xi|^{-1}\,||_{L^\infty}^2 ||\nabla^2\xi||_{L^p}||u_\epsilon||_{C^0}$
- Terms with L^p norm bounded by $||\,|\nabla\xi|^{-1}\,||_{L^\infty}||\nabla u_\epsilon||_{L^p}$.

One can check this estimate by enumerating the terms, but we point out that these estimates are exactly what one expects from a solution to $i\partial_j\xi q^{jl} = u^l$ if one thinks of q^{jl} as schematically given by "$q\approx\frac{u}{\nabla\xi}$" and applies the quotient rule formally. Note that, by the Sobolev embedding theorem, we can also control $||u_\epsilon||_{C^0}$ with $||u_\epsilon||_{W^{1,p}}$, so that only the $W^{1,p}$ norm of u_ϵ needs to appear in the estimate.

The upshot of these estimates is that we can now bound

$$||Q_1^{jl}||_{C^0(\mathbb{T}^3)} \lesssim \epsilon^s||u||_{W^{s,p}} + \frac{1}{\lambda}||\partial_j q^{jl}||_{L^p}$$

$$\lesssim \epsilon^s||u||_{W^{s,p}} + \frac{||\,|\nabla\xi|^{-1}\,||_{L^\infty}}{\lambda}\left(1 + ||\,|\nabla\xi|^{-1}\,||_{L^\infty}||\nabla^2\xi||_{L^p}\right)||u_\epsilon||_{W^{1,p}}$$

$$\lesssim \epsilon^s||u||_{W^{s,p}} + \frac{||\,|\nabla\xi|^{-1}\,||_{L^\infty}}{\lambda}\left(1 + ||\,|\nabla\xi|^{-1}||_{L^\infty}||\nabla^2\xi||_{L^p}\right)\epsilon^{s-1}||u||_{W^{s,p}}.$$

It is now clear that, regardless of s, the optimal choice of ϵ, which balances the two terms, is

$$\epsilon = \frac{||\,|\nabla\xi|^{-1}\,||_{L^\infty}}{\lambda}\left(1 + ||\,|\nabla\xi|^{-1}\,||_{L^\infty}||\nabla^2\xi||_{L^p}\right).$$

As expected, this choice of ϵ ensures that u_ϵ can only oscillate at a scale coarser than the scale at which the phase function $\lambda\xi$ oscillates.[8] With the above choice of ϵ, the bounds stated in the theorem apply to Q_1^{jl}. By the same considerations, the parametrix $\frac{1}{i\lambda}e^{i\lambda\xi}q_\epsilon^{jl}$ can also be controlled in C^0 by $\frac{||\,|\nabla\xi|^{-1}\,||_{L^\infty}}{\lambda}||u_\epsilon||_{W^{1,p}}$, and therefore obeys the estimates stated in the theorem, which concludes the proof. $\qquad\square$

6.3   Higher Order Parametrix Expansion

Later on in the proof, we will have to modify the argument used in the proof of Proposition (6.1) for solving Equation (35). Namely, we will construct a solution to

$$\partial_j Q^{jl} = e^{i\lambda\xi(x)}u^l \tag{39}$$

---

[8]We remark that when $s = 1$, the result can be obtained without any mollification ($\epsilon = 0$).

by taking a higher order expansion of the parametrix

$$Q^{jl} = \tilde{Q}^{jl}_{(D)} + Q^{jl}_{(D)} \tag{40}$$

$$\tilde{Q}^{jl}_{(D)} = \sum_{(k)=1}^{D} e^{i\lambda\xi} \frac{q^{jl}_{(k)}}{\lambda^k} \tag{41}$$

where each $q_{(k)}$ solves a linear equation

$$i\partial_j\xi q^{jl}_{(1)} = u^l \tag{42}$$

$$i\partial_j\xi q^{jl}_{(k)} = u^l_{(k)} \tag{43}$$

$$u^l_{(k)} = -\partial_j q^{jl}_{(k-1)} \qquad 1 < k \le D+1 \tag{44}$$

using the linear map

$$q^{jl} = q^{jl}(\nabla\xi)[u] \tag{45}$$

defined in (38). Then Q_D will be used to eliminate the error

$$\partial_j Q^{jl}_D = e^{i\lambda\xi} \frac{u^l_{(D+1)}}{\lambda^D}. \tag{46}$$

The reason we will need a higher order expansion for constructing Hölder continuous solutions is that the parameter λ will be large but limited in size. In fact u will have "frequency" Ξ where Ξ is basically the frequency λ chosen in the previous stage of the iteration, while we will have $\lambda \sim \Xi^{1+\eta}$ for some small $\eta > 0$. In this case, one can only obtain an accurate approximate solution from the parametrix after expanding the parametrix to high order.

6.4 An Inverse for Divergence

One way to prove Lemma (6.1) is to define the following operator, whose symbol in frequency space is exactly the map $q^{jl}(\nabla\xi)[u]$ defined in Line (38) of the proof of Proposition (6.1).

Proposition 6.2 (Elliptic Estimates). *There exists a linear operator $\mathcal{R}^{jl}$ such that for all vector fields $U \in C^\infty(\mathbb{T}^3)$ we have*

$$\partial_j \mathcal{R}^{jl}[U] = \mathcal{P}[U]^l$$

where $\mathcal{P}$ denotes the projection to integral 0 vector fields.
Furthermore, $\mathcal{R}$ satisfies the bounds

$$||\nabla^{k+1}\mathcal{R}[U]||_{L^4(\mathbb{T}^3)} \le C_k ||\nabla^k U||_{L^4(\mathbb{T}^3)} \qquad k \ge 0 \tag{47}$$

and gives a solution with integral 0

$$\int \mathcal{R}^{jl}[U]dx = 0. \tag{48}$$

28

Remark. Here we only state estimates for L^4 norms rather than general L^p norms, since the L^4 estimate suffices for our proof.

Proof. We define $\mathcal{R}$ as a sum

$$\mathcal{R}[U] = \mathcal{R}_1[U] + \mathcal{R}_2[U] \tag{49}$$

using the Helmholtz decomposition of U into its divergence free and curl free parts.

$$U^l = \partial^l \Delta^{-1} \partial_i U^i + (U^l - \partial^l \Delta^{-1} \partial_i U^i) \tag{50}$$

$$= \partial_j [\Delta^{-1} \partial_i U^i \delta^{jl}] + (U^l - \partial^l \Delta^{-1} \partial_i U^i) \tag{51}$$

$$= \partial_j [\partial_i (\Delta^{-1} \mathcal{P} U^i) \delta^{jl}] + (U^l - \partial^l \Delta^{-1} \partial_i U^i) \tag{52}$$

$$= \partial_j [\mathcal{R}_1^{jl}[U]] + \mathcal{H} U^l \tag{53}$$

Observe that $\mathcal{H} U^l$ is divergence free, so to define $\mathcal{R}_2$, it suffices to choose

$$\mathcal{R}_2^{jl}[U] = \partial^j [\Delta^{-1} \mathcal{P} \mathcal{H} U^l] + \partial^l [\Delta^{-1} \mathcal{P} \mathcal{H} U^j] \tag{54}$$

so that

$$\partial_j \mathcal{R}_2^{jl}[U] = \mathcal{P} \mathcal{H} U^l \tag{55}$$

and

$$\partial_j [\mathcal{R}_1^{jl}[U] + \mathcal{R}_2^{jl}[U]] = \partial_j [\Delta^{-1} \partial_i U^i \delta^{jl}] + \mathcal{P}(U^l - \partial^l \Delta^{-1} \partial_i U^i) \tag{56}$$

$$= \mathcal{P} U^l. \tag{57}$$

For $k = 0$, the estimate (47) reads

$$\|\nabla \mathcal{R}[U]\|_{L^4(\mathbb{T}^3)} \leq C \|U\|_{L^4(\mathbb{T}^3)}. \tag{58}$$

This inequality is a consequence of the local Calderón Zygmund estimate

$$\|D^2 f\|_{L^4(B)} \leq C(\|f\|_{L^4(2B)} + \|\Delta f\|_{L^4(2B)}) \tag{59}$$

applied to $f^l = \Delta^{-1} \mathcal{P} U^l$ and taking any ball $B \subseteq \mathbb{R}^3$ containing an entire periodic box $\mathbb{T}^3 \subseteq (B/\mathbb{Z}^3)$. In this case, because $f^l = \Delta^{-1} \mathcal{P} U^l$ has integral 0, we are able to replace the term

$$\|f^l\|_{L^4} = \|\Delta^{-1} \mathcal{P} U\|_{L^4}$$

by

$$\|\Delta^{-1} \mathcal{P} U\|_{L^4(2B)} \leq C \|\Delta^{-1} \mathcal{P} U\|_{L^4(\mathbb{T}^3)} \tag{60}$$

$$\leq C \|U\|_{L^4(\mathbb{T}^3)} \tag{61}$$

which can be quickly verified using, for example, Littlewood-Paley theory.

This estimate finishes the proof of (47) for $k = 0$. For $k > 0$, the bound follows from the fact that $\mathcal{R}$ commutes with spatial derivatives. $\qquad \square$

7 Constructing the Correction

7.1 Transportation of the Phase Functions

We begin the main construction by considering the Transport term. Let us fix a solution (v, p, R) to the Euler-Reynolds equations and consider a correction $v_1 = v + V$, $p_1 = p + P$. We will think of v as an approximation to the "coarse scale velocity" since the solution ultimately achieved by the process will resemble v at a sufficiently coarse scale.

We must eliminate the Transport term by solving the equation

$$\partial_j Q_T^{jl} = \partial_t V^l + \partial_j (v^j V^l)$$

with a stress Q_T^{jl} of size $|Q| < |R|$. Ideally, the stress Q can be made arbitrarily small by choosing λ sufficiently large.

Recall that the correction V^l takes the form of a superposition of waves $V^l = \sum_I V_I^l$, where each individual wave V_I^l is an oscillatory, divergence free vector field of the form

$$V_I^l = e^{i\lambda \xi_I} \tilde{v}_I^l.$$

Before we discuss how to handle the Stress term, we must leave the amplitude $\tilde{v}_I^l$ unspecified, however we will insist that its space-time support be such that, among other things, $|\nabla \xi_I|$ remains bounded from 0 so that the oscillatory estimate can be applied. Substituting for V and differentiating naively gives

$$\partial_t V^l + \partial_j (v^j V^l) = (\partial_t + v^j \partial_j) V^l$$
$$= \sum_I e^{i\lambda \xi_I} [(i\lambda)(\partial_t + v^j \partial_j) \xi_I \tilde{v}_I^l + (\partial_t + v^j \partial_j) \tilde{v}_I^l].$$

As we are able to gain at most a power of λ from solving the divergence equation, it seems that unless ξ_I obeys the transport equation $(\partial_t + v^j \partial_j) \xi_I = 0$, we will not be able to adequately control the size of the solution Q_T^{jl} – formally, the data is at least of size $\lambda |\tilde{v}_I| \sim \lambda |R|^{1/2}$ giving a solution Q that is at best size $|R|^{1/2}$. This impression is almost correct, however, one cannot afford to allow ξ_I to inherit the fine scale oscillations of the coarse scale velocity v^j. Instead, it is natural to limit the frequency of the Transport term, by using a mollifier $v_\epsilon^j = \eta_{\epsilon_v} * v^j$, and letting ϵ_v be a parameter that must be chosen sufficiently small.[9]

We will refer to v_ϵ as the **coarse scale velocity**.

So in fact we treat the Transport term by first decomposing

$$\partial_t V^l + \partial_j (v^j V^l) = \partial_t V^l + v_\epsilon^j \partial_j V^l + \partial_j ((v^j - v_\epsilon^j) V^l)$$

---

[9]To produce continuous solutions the mollifier η_{ϵ_v} may be in both space and time variables. For Hölder continuous solutions, we will only mollify v in the spatial variables.

This mollification introduces a term

$$(v^j - v_\epsilon^j)V^l + V^j(v^l - v_\epsilon^l)$$

which will constitute part of the new stress R_1^{jl}, and the parameters ϵ_v and λ will later be chosen sufficiently small and large respectively so that this term is acceptably small.

The Transport term can now be written as

$$\partial_t V^l + v_\epsilon^j \partial_j V^l = \sum_I e^{i\lambda\xi_I}[(i\lambda)(\partial_t + v_\epsilon^j \partial_j)\xi_I \tilde{v}_I^l + (\partial_t + v_\epsilon^j \partial_j)\tilde{v}_I^l].$$

We will require that all the phase functions obey the same transport equation:

$$(\partial_t + v_\epsilon^j \partial_j)\xi_I = 0 \tag{62}$$

$$\xi_I(t(I), x) = \hat{\xi}_I \tag{63}$$

Therefore, although we will choose linear initial data $\hat{\xi}_I$ for the phase functions ξ_I, they will immediately cease to be linear, and will only remain useful in the oscillatory estimates for a short period of time $|t - l(I)| \leq \tau$ depending on v. We are therefore forced to embed appropriate time cutoff functions into the definition of the amplitudes $\tilde{v}_I^l$. These cutoff functions must restrict the lifespan of V_I so that the following requirements remain fulfilled:

- **Nondegeneracy**: We require $|\nabla\xi_I|$ to remain bounded from 0.

- **Higher Derivative Bounds**: We will also require that $D^2\xi_I$ remains bounded (uniformly in I), because the oscillatory estimate requires us to differentiate the amplitude of the data.

The price we pay for introducing a time cutoff function is that the time cutoff itself will be differentiated in the Transport term, and the shorter the lifespan, the larger the induced stress. Therefore, the time parameter τ governing the lifespan of V_I must be chosen to achieve the above objectives before the parameter λ is chosen to ensure that the resulting stress is small.

It is interesting to reflect on how some approximation to this phase transport requirement can be seen in previous constructions of solutions to the Euler equation which fail to conserve energy, including Shnirelman's use of "modulated Kolmogorov flow" in [Shn97], the late stages of the solutions to Tartar's wave cone used in the argument [DLS09], and the phase modulations in the argument [DLS13].

We have now explained the constraint that the Transport term imposes on the phase functions. We cannot begin to make estimates on the solution until we have specified the amplitudes v_I of the correction, and these cannot be specified until we describe how to eliminate the Stress term

31

$Q_S^{jl} = \sum_I (V_I^j \bar{V}_I^l) + P_0 \delta^{jl} + \partial_j R^{jl}$. However, before we can resolve the Stress term, we must first explain how the High-High Interference term is dealt with, since eliminating these terms will involve imposing an additional constraint on the amplitudes v_I.

7.2   The High-High Interference Problem and Beltrami Flows

The High-High Interference term is given by

$$\partial_j Q_H^{jl} = \sum_{J \neq \bar{I}} \partial_j (V_I V_J)^{jl} \tag{64}$$

where each V_I is an oscillatory wave of the form $V_I = e^{i\lambda \xi_I} \tilde{v}_I$, and the condition that $I \neq \bar{J}$ will ensure that $\nabla \xi_J$ is separated from $-\nabla \xi_I$, so that the symmetric product $(V_I V_J)^{jl} = \frac{1}{2} e^{i\lambda(\xi_I + \xi_J)} (\tilde{v}_I^j \tilde{v}_J^l + \tilde{v}_I^j \tilde{v}_J^l)$ genuinely has high frequency after we have proven an upper bound on $|\nabla(\xi_I + \xi_J)|^{-1}$.

Interference terms such as these are typical in even the simplest examples of convex integration and other related constructions such as the construction of nowhere differentiable functions. There must be a way to ensure that the correction chosen to achieve a certain objective in a region Ω does not interfere with the correction chosen in an overlapping region Ω', otherwise there is a danger of "stepping on your own foot" during the construction. The device introduced in [DLS13] which overcomes this difficulty for the Euler equations is to look at Beltrami flows. These are eigenfunctions of the curl operator ($\nabla \times v = \mu v$), which are automatically stationary solutions to the Euler equations.[10]

Here we will impose a microlocal, Beltrami flow condition. To be specific, recall that the individual waves take the form

$$V_I = \frac{1}{\lambda} \nabla \times (e^{i\xi_I} w_I) = e^{i\lambda \xi_I} (v_I + \frac{\nabla \times w_I}{\lambda})$$

where w_I is a pointwise solution to the linear equation $(i\nabla \xi_I) \times w_I = v_I$, so that the amplitudes v_I are required to point in a direction perpendicular to the direction of oscillation $\nabla \xi_I$, like any high frequency, divergence free plane wave should. The high frequency analogue of the Beltrami flow condition $\nabla \times v = \mu v$ is the pointwise condition $(i\nabla \xi_I) \times v_I = \mu v_I$, where the eigenvalue μ must be one of $\mu = \pm |\nabla \xi_I|$ since those are the eigenvalues of $(i\nabla \xi_I) \times$ when viewed as an operator on $\langle \nabla \xi_I \rangle^{\perp}$.

We will therefore require that

$$\nabla \xi_I \cdot v_I = 0$$
$$(i\nabla \xi_I) \times v_I = |\nabla \xi_I| v_I$$

---

[10]More generally, a Beltrami flow can be of the form $\nabla \times v = \mu(x) v$ for an arbitrary function $\mu(x)$.

holds pointwise, and choose

$$w_I = \frac{(i\nabla\xi_I)}{|\nabla\xi_I|^2} \times v_I = \frac{1}{|\nabla\xi_I|}v_I.$$

If we untangle this requirement in terms of the real and imaginary parts $a_I^l, b_I^l \in \mathbb{R}^3$ of $v_I^l = a_I^l + ib_I^l$, then the Beltrami flow condition can be phrased as:

$$\nabla\xi_I \cdot a_I = 0, \quad b_I = \frac{(\nabla\xi_I)}{|\nabla\xi_I|} \times a_I$$

or equivalently

$$\nabla\xi_I \cdot b_I = 0, \quad a_I = -\frac{(\nabla\xi_I)}{|\nabla\xi_I|} \times b_I.$$

Therefore, our final remaining degree of freedom in the choice of V_I is, without loss of generality, the imaginary part of its amplitude, b_I, which itself is required to be perpendicular to $\nabla\xi_I$. We will specify this amplitude, its support in space-time, along with the index set I in the following section. For now, we explain why the choice of approximate Beltrami flows helps to estimate the High-High Interference term (essentially, we will go through the proof that Beltrami flows are stationary solutions to Euler).

We rewrite (64) as

$$\partial_j Q_H^{jl} = \frac{1}{2}\sum_{J \neq \bar{I}} \partial_j(V_I^j V_J^l + V_J^j V_I^l) \tag{65}$$

and use the divergence free condition to write

$$\partial_j Q_H^{jl} = \frac{1}{2}\sum_{J \neq \bar{I}} V_I^j \partial_j V_J^l + V_J^j \partial_j V_I^l. \tag{66}$$

Let us fix a pair I, J, and permute the indices

$$V_I^j \partial_j V_J^l + V_J^j \partial_j V_I^l = (V_I)_j[\partial^j V_J^l - \partial^l V_J^j] + (V_J)_j[\partial^j V_I^l - \partial^l V_I^j] \tag{67}$$
$$+ (V_I)_j\partial^l V_J^j + (V_J)_j\partial^l V_I^j. \tag{68}$$

The second of these terms is actually a gradient

$$(V_I)_j\partial^l V_J^j + (V_J)_j\partial^l V_I^j = \partial^l(V_I \cdot V_J), \tag{69}$$

and it will be absorbed into the pressure by a term

$$P = P_0 + \sum_{J \neq \bar{I}} P_{I,J} \tag{70}$$

$$P_{I,J} = -\frac{(V_I \cdot V_J)}{2}. \tag{71}$$

33

We now are reduced to solving

$$\partial_j Q^{jl}_{H,(IJ)} = (V_I)_j[\partial^j V^l_J - \partial^l V^j_J] + (V_J)_j[\partial^j V^l_I - \partial^l V^j_I]. \qquad (72)$$

In three dimensions, a special observation can be made using the alternating structure of the right-hand side. One way to see it is to recall some notation. Let us recall the volume element ϵ^{abc} which is the fully antisymmetric tensor defined so that the trilinear form $\epsilon^{abc} u_a v_b w_c$ gives the signed volume of the parallelogram $u \wedge v \wedge w$. We recall the basic definitions $(u \times v)^a = \epsilon^{abc} u_b v_c$ and $(\nabla \times v)^a = \epsilon^{abc} \partial_b v_c$. Using these definitions and the elementary identity[11]

$$\epsilon^{abc}\epsilon_{cfg} = \delta^a_f \delta^b_g - \delta^a_g \delta^b_f$$

we establish the calculus identity

$$\begin{aligned}
(V_I)_j[\partial^j V^l_J - \partial^l V^j_J] &+ (V_J)_j[\partial^j V^l_I - \partial^l V^j_I] \\
&= -V_I \times (\nabla \times V_J) - V_J \times (\nabla \times V_I).
\end{aligned} \qquad (73)$$

At this stage, we can see why curl eigenfunctions are helpful. Namely, if we were using eigenfunctions of curl

$$\nabla \times V_I = \lambda_I V_I \qquad (74)$$

$$\nabla \times V_J = \lambda_J V_J \qquad (75)$$

$$\lambda_I = \lambda_J \qquad (76)$$

with the same eigenvalue, then their sum would be a curl eigenfunction and (73) would be identically 0. The "microlocal Beltrami-flow" condition says that v_I is an eigenfunction for the symbol of curl, namely

$$(i\nabla\xi_I) \times v_I = |\nabla\xi_I|v_I. \qquad (77)$$

In this way, each V_I looks like a curl eigenfunction to leading order in λ.

---

[11]This identity gives two different expressions for the inner product on the second exterior power of $\Lambda^2(\mathbb{R}^3)$. It can be proven by testing against components of an orthonormal frame, and the proof can be accelerated by observing that these tensors are both antisymmetric in (ab) and (fg).

We now express (72) more completely as

$$
\begin{aligned}
\partial_j Q^{jl}_{H,(IJ)} &= -\lambda e^{i\lambda(\xi_I+\xi_J)} \left(\tilde{v}_I \times [(i\nabla\xi_J) \times \tilde{v}_J] + \tilde{v}_J \times [(i\nabla\xi_I) \times \tilde{v}_I] \right) \\
&\quad - e^{i\lambda(\xi_I+\xi_J)} \left(\tilde{v}_I \times (\nabla \times \tilde{v}_J) + \tilde{v}_J \times (\nabla \times \tilde{v}_I) \right) \\
&= -\lambda e^{i\lambda(\xi_I+\xi_J)} \left(v_I \times [(i\nabla\xi_J) \times v_J] + v_J \times [(i\nabla\xi_I) \times v_I] \right) \\
&\quad -e^{i\lambda(\xi_I+\xi_J)}((\nabla \times w_I) \times [(i\nabla\xi_J) \times \tilde{v}_J] + (\nabla \times w_J) \times [(i\nabla\xi_I) \times \tilde{v}_I]) \\
&\quad -e^{i\lambda(\xi_I+\xi_J)}(v_I \times [(i\nabla\xi_J) \times (\nabla \times w_J)] + v_J \times [(i\nabla\xi_I) \times (\nabla \times w_I)]) \\
&\quad -e^{i\lambda(\xi_I+\xi_J)}(\tilde{v}_I \times (\nabla \times \tilde{v}_J) + \tilde{v}_J \times (\nabla \times \tilde{v}_I))
\end{aligned}
$$

$$(78)$$

$$
\begin{aligned}
&= -\lambda e^{i\lambda(\xi_I+\xi_J)} \left(v_I \times [(i\nabla\xi_J) \times v_J] + v_J \times [(i\nabla\xi_I) \times v_I] \right) \\
&\quad + \text{ Lower Order Terms.}
\end{aligned}
$$

$$(79)$$

Using the condition (77), we add

$$
\lambda e^{i\lambda(\xi_I+\xi_J)}(v_I \times v_J + v_J \times v_I) = 0
$$

to rewrite the High-High Interference term (79) as

$$
\partial_j Q^{jl}_{H,(IJ)} - -\lambda e^{i\lambda(\xi_I+\xi_J)} \left(v_I \times [(|\nabla\xi_J|-1)v_J] + v_J \times [(|\nabla\xi_I|-1) \times v_I] \right) \tag{80}
$$
$$
+ \text{ Lower Order Terms.}
$$

Our strategy for ensuring this term can be controlled involves imposing the following additional conditions on the phase functions:

- **Synchronized Periods**: We require that $|\nabla\xi_I| \approx 1 \approx |\nabla\xi_J|$ for all I, J.

- **Angle Separation**: We require $|\nabla\xi_I + \nabla\xi_J|$ to remain bounded from 0 for $J \neq \bar{I}$ so that the parametrix can be used. Since the $\nabla\xi_I$ and $\nabla\xi_J$ both have absolute value essentially one, this lower bound is equivalent to the angles between the gradients being well-separated.

These properties will be guaranteed by taking initial values

$$
\xi_I(t(I), x) = \hat{\xi}_I \tag{81}
$$

for the phase functions which have angle-separated gradients, and such that $|\nabla\hat{\xi}_I| = 1$ everywhere. We then use the time cutoff function in the amplitude to make sure that the common lifespan of the V_I is sufficiently short so that the equal periodicity condition is maintained. This lifespan will depend on the given v, as the gradient of v_ϵ enters into the transport equations for the phase gradients.

From our preceding discussion, we have established that the final remaining degree of freedom in constructing the correction goes into choosing

b_I^l, the imaginary part of the amplitude of v_I^l, as well as precisely specifying the phase functions. We know that the amplitudes b_I will involve cutoff functions in space and time chosen so that the phase functions ξ_I can exist on the support of b_I without any critical points, so that the derivatives of the ξ_I can remain bounded, and also so that the building blocks will still approximate Beltrami flows. Up to the choice of some parameters, the construction will be completely specified in the following section, where the amplitudes b_I are chosen to make the contribution of the Stress term small.

7.3 Eliminating the Stress

Here we discuss how to construct the corrections V_I, P_0 in such a way that the Stress term

$$Q_S^{jl} = \sum_I (V_I^j \bar{V}_I^l) + P_0 \delta^{jl} + R^{jl} \qquad (82)$$

can be reduced to a new stress Q_S^{jl} much smaller than R^{jl} in C^0 after appropriate choices of V_I and P_0. In the process, we will describe the initial data (63) for the phase functions and also the set $\mathcal{I}$ by which the corrections V_I are indexed. A brief summary of the results of this section are summarized by the proposition in Section (7.3.7).

In this part of the argument, we will take a different approach than the one taken in [DLS13]. Their approach is based on Carathéodory's theorem for convex hulls, and does not seem to generalize readily to the setting of nonlinear phase functions.

Unlike the Transport term, the High-Low Interaction term and the High-High Interference terms, the Stress term (82) does not require us to solve a divergence equation. Rather, eliminating this term is essentially an algebraic problem which can be thought of as calculating an elaborate "square root."

We will choose $V = \sum_I V_I$ and P_0 to ensure that

$$\sum_I (V_I^j \bar{V}_I^l) + P_0 \delta^{jl} + R^{jl} \approx 0 \qquad (83)$$

is sufficiently small pointwise.

Roughly speaking, making the error (83) small pointwise will give the amplitudes of V_I size $|V_I| \lesssim |R|^{1/2}$. We will need to measure the size of the derivatives of V_I, so we will attempt to ensure that the terms V_I are balanced in size $|V_I| \sim |R|^{1/2}$ where they are supported, since a derivative $\partial R^{1/2}$ can be quite large when $|R|$ is small.

7.3.1 The Approximate Stress Equation

Before starting the construction, we observe that one only has a hope of solving this equation if $P_0 \delta^{jl} + R^{jl}$ is negative definite. Although we cannot fully eliminate R^{jl} by adding a multiple of the metric δ^{jl}, we can at least ensure that the sum $P_0 \delta^{jl} + R^{jl}$ is negative definite, and is also not significantly larger than R^{jl} itself by choosing P_0 of size $|P_0| \sim |R|$.

The other issue we must take into account is that if we solve the equation $\sum_I (V_I^j \bar{V}_I^l) + P_0 \delta^{jl} + R^{jl} = 0$ pointwise, then the individual waves V_I inherit the fine scale behavior of R^{jl} (which is inconsistent with the basic philosophy of the construction), and their higher derivatives will therefore not obey good estimates. A similar issue arose while considering the Transport term.

To deal with the latter issue, we will first replace R by a version $R_\epsilon = \eta_{\epsilon_R} *_{x,t} R$ which is mollified in both space and time variables. The scale ϵ_R of this mollification will be chosen later in the argument so that the error $R - R_\epsilon$ is acceptably small. Actually, in order to achieve sharp regularity in the main theorem, it will be necessary to do something more complicated than a simple-minded space-time mollification.

Finally, observe that, without loss of generality, we can assume that R_ϵ is trace free by absorbing the trace part $(R_\epsilon^{jl} \delta_{jl}) \frac{\delta^{jl}}{n}$ into the pressure. More specifically, we choose a correction of the form

$$P_0 = -\frac{e(t)}{n} - \frac{(R_\epsilon^{jl} \delta_{jl})}{n}, \qquad n = 3 \tag{84}$$

and with this choice the equation

$$\sum_I (V_I^j \bar{V}_I^l) = -P_0 \delta^{jl} - R_\epsilon^{jl} \tag{85}$$

takes the form

$$\sum_I (V_I^j \bar{V}_I^l) = e(t) \frac{\delta^{jl}}{n} - \mathring{R}_\epsilon^{jl} \tag{86}$$

where the tensor

$$\mathring{R}_\epsilon^{jl} = R_\epsilon^{jl} - (R_\epsilon^{jl} \delta_{jl}) \frac{\delta^{jl}}{n}$$

is the trace free part of R_ϵ.

Here $e(t)$ is a non-negative function which must satisfy certain restrictions such as those of Lemma (5.1). For example, we impose a lower bound (18) on $e(t)$ to make sure that the right-hand side of (86) is, among other things, non-negative definite.

Prescribing the trace $e(t)$ in Equation (86) gives us precise control over the energy

$$\int |V|^2 dx = \int V^j V^l \delta_{jl} dx$$

added into the system by the correction V. Namely, since the components V_I and V_J oscillate rapidly in different directions for $J \neq \bar{I}$, they are almost orthogonal

$$\int V_I \cdot V_J dx \approx 0 \qquad J \neq \bar{I}$$

leading to an approximate energy increase of

$$\int |V|^2 dx = \sum_{I,J} \int V_I \cdot V_J dx \tag{87}$$

$$\approx \int \sum_I V_I \cdot \bar{V}_I dx \tag{88}$$

$$\approx \int_{\mathbb{T}^3} e dx \tag{89}$$

if we can solve Equation (86). Therefore, in terms of dimensional analysis, the function e, much like p and R, should be regarded as having the dimensions of an energy density (or $\frac{length^2}{time^2}$).

Of course, we cannot solve Equation (86) exactly, because we have also required that each V_I be divergence free. However, as we explain in the following section, Equation (86) can be solved approximately when the oscillation parameter λ is sufficiently large.

7.3.2   The Stress Equation and the Initial Phase Directions

Let us recall that each V_I can be represented as

$$V_I = e^{i\lambda \xi_I} \tilde{v}_I$$

$$= \nabla \times (\frac{e^{i\lambda \xi_I}}{\lambda} w_I)$$

$$= e^{i\lambda \xi_I} (v_I + \frac{\nabla \times w_I}{\lambda}).$$

When λ is large, the v_I term dominates the $\frac{\nabla \times w_I}{\lambda}$ term, and we have an approximation

$$V_I \approx e^{i\lambda \xi_I} v_I.$$

We still have freedom to choose the amplitude v_I, so instead of solving Equation (86) exactly, we will solve the following equation:

The Stress Equation

$$\sum_I v_I^j \bar{v}_I^l = e \frac{\delta^{jl}}{n} - \mathring{R}_\epsilon^{jl} \tag{90}$$

Solving (90) will leave an error term

$$Q_S^{jl} = \sum_I \frac{(\nabla \times w_I)^j \bar{v}_I^l}{\lambda} + \sum_I \frac{v_I^j (\nabla \times \bar{w}_I)^l}{\lambda}$$

$$+ \sum_I \frac{(\nabla \times w_I)^j (\nabla \times \bar{w}_I)^l}{\lambda^2}, \tag{91}$$

which composes part of the new stress R_1^{jl}.

In Section (7.2), we introduced the requirement that the complex amplitude v_I must satisfy the microlocal Beltrami flow condition $(i\nabla\xi_I) \times v_I = |\nabla\xi_I|v_I$, and assuming this requirement was satisfied, we defined $w_I = \frac{(i\nabla\xi_I)}{|\nabla\xi_I|^2} \times v_I = \frac{1}{|\nabla\xi_I|}v_I$. If $v_I = a_I + ib_I$ is expanded into its real and imaginary parts, then the Beltrami flow condition can be written $b_I = \frac{(i\nabla\xi_I)}{|\nabla\xi_I|} \times a_I$.

The imaginary part b_I is our last remaining degree of freedom, but it is still constrained by the requirement that

$$\nabla\xi_I \cdot b_I = \partial_l \xi_I b_I^l = 0$$

expressing the condition that the waves be divergence free. The real part a_I defined by

$$a_I = -\frac{(i\nabla\xi_I)}{|\nabla\xi_I|} \times b_I \tag{92}$$

satisfies the same requirement, since it is a rotation of b_I within the plane orthogonal to $\nabla\xi_I$ by an angle of $\frac{\pi}{2}$.

By expanding each v_I in terms of real and imaginary parts, Equation (90) becomes

$$\sum_I (a_I^j a_I^l + b_I^j b_I^l) = e\frac{\delta^{jl}}{n} - \mathring{R}_\epsilon^{jl}.$$

No matter how b_I is chosen, as long as b_I is indeed orthogonal to $\nabla\xi_I$, the bilinear form that arises on the left-hand side is given by

$$a_I^j a_I^l + b_I^j b_I^l = |b_I|^2(\delta^{jl} - \frac{\partial^j \xi_I \partial^l \xi_I}{|\nabla\xi_I|^2})$$

where $\delta^{jl} - \frac{\partial^j \xi_I \partial^l \xi_I}{|\nabla\xi_I|^2}$ is the orthogonal projection of the inner product δ^{jl} to the plane $\langle\nabla\xi_I\rangle^\perp$. This fact follows from Equation (92), which implies that a_I is orthogonal to b_I and has the same absolute value. One can also check this identity by showing that the form $v_I^j \bar{v}_I^l$ restricted to the plane $\langle\nabla\xi_I\rangle^\perp$ is invariant under any rotation within the plane, and is determined up to a constant by this property.

Using these observations we can write Equation (90) in the form

$$\sum_I |b_I|^2(\delta^{jl} - \frac{\partial^j \xi_I \partial^l \xi_I}{|\nabla\xi_I|^2}) = G^{jl}, \tag{93}$$

for a tensor

$$G^{jl} = e\frac{\delta^{jl}}{n} - \mathring{R}_\epsilon^{jl} \tag{94}$$

39

which is positive definite as long as

$$e \geq K|R_\epsilon| \tag{95}$$

for a sufficiently large constant K. The constant K in (95) is an absolute constant which we will wait until later to specify. Regarding the choice of e, the reader should keep in mind that we expect an estimate of the type $|V_I| \approx |b_I| \lesssim |R|^{1/2}$. Furthermore, obtaining bounds on the derivatives of V_I will basically require us to differentiate Equation (93), which will require lower bounds on $|b_I|$ as well.

With the stress equation in the form (93), we can now explain how the phase functions ξ_I and the amplitudes b_I are constructed.

7.3.3 The Index Set, the Cutoffs and the Phase Functions

We have established already that the phase functions ξ_I must satisfy a transport equation

$$(\partial_t + v_\epsilon^j \partial_j)\xi_I = 0.$$

What remains is to specify their data. Each phase function will be given an initial condition at time $t(I)$. To assure that equation (93) may be satisfied, we require from the the phase functions that the tensors

$$\delta^{jl} - \frac{\partial^j \xi_I \partial^l \xi_I}{|\nabla \xi_I|^2}$$

must span the space $\mathcal{S}$ of symmetric, $(2,0)$ tensors at every point. Therefore, at each point in $\mathbb{R} \times \mathbb{T}^3$ we will require at least 6 phase functions to solve (93) – or 12, rather, since the negative of each phase function will also be included. The phase functions will only be defined on some open set $\Omega_I(t) \subseteq \mathbb{T}^3$, where Ω_I is the image of the parameterization $(\Phi_I(x,t), t) : \Omega_0 \times \mathbb{R} \to \mathbb{T}^3 \times \mathbb{R}$, which moves under the flow of v_ϵ.

The amplitude function will be equipped with two cutoff functions:

$$b_I^l = \eta((t - t(I))/\tau)\psi_I(x,t)b_{I,1}^l = \eta_I \psi_I b_{I,1}^l$$

The cutoff ψ_I localizes to Ω_I, so that we will not have to worry that the phase function fails to be defined globally. In order to be compatible with the transport of ξ_I, we will also have

$$(\partial_t + v_\epsilon^j \partial_j)\psi_I = 0,$$

so we only have to specify the initial values for both ξ_I and ψ_I. The cutoff $\eta((t - t(I))/\tau)$ limits the lifespan of V_I to a time period less than or equal to τ, where τ is a small parameter that remains to be chosen.

To continue, we will have to explain the set $\mathcal{I}$ to which the indices I belong. The purpose of the index I is to record both the location of V_I in space-time, and the direction $\nabla \xi_I$ in which V_I oscillates. The strategy is the following: at time 0, we cover the torus $\mathbb{T}^3$ with 8 charts, and on each of these charts we place 12 phase functions ξ_I so that the six directions determined by the ξ_I allow us to solve (93) for an arbitrary right-hand side within that chart. We must cut off these amplitudes after a very short time interval, at which point the phase functions must be replaced by another group of phase functions whose amplitudes will also be supported on 8 charts. To avoid interference in the High-high interactions, we also must make sure that any pair V_I and V_J who share support in space-time oscillate in well-separated directions. The cutoff functions are chosen as part of a partition of unity which allows us to patch together local solutions of the quadratic Equation (93), and such a choice is consistent with the transport equation.

The Index Set as a Graph and Some Notation With this rough description in place, we choose the index set $\mathcal{I} = (\mathbb{Z}/(2\mathbb{Z}))^3 \times \mathbb{Z} \times F$. The F coordinate is meant to specify the direction $d\xi_I$ in which V_I oscillates, so F will have 12 elements; to be concrete, let $\varphi = (1 + \sqrt{5})/2$ be the golden ratio and set

$$F = \left\{ \pm\frac{(0, 1, \pm\varphi)}{\sqrt{1 + \varphi^2}}, \pm\frac{(1, \pm\varphi, 0)}{\sqrt{1 + \varphi^2}}, \pm\frac{(\pm\varphi, 0, 1)}{\sqrt{1 + \varphi^2}} \right\}$$

to be the set of normal vectors to the faces of a dodecahedron in $\mathbb{R}^3$, which themselves form the vertices of an icosahedron. It will also be useful to define the projective dodecahedron

$$\mathbb{F} = F/\langle\pm 1\rangle = \left\{ \left[\frac{(0, 1, \pm\varphi)}{\sqrt{1 + \varphi^2}}\right], \left[\frac{(1, \pm\varphi, 0)}{\sqrt{1 + \varphi^2}}\right], \left[\frac{(\pm\varphi, 0, 1)}{\sqrt{1 + \varphi^2}}\right] \right\}.$$

The advantage of doing so is that $\mathbb{F}$ inherits the action of the icosahedral group, but accounts for each of the $|\mathbb{F}| = 6$ directions in F exactly once.

The discrete coordinate $k(I) = (\kappa, k_4) = (\kappa_1, \kappa_2, \kappa_3, k_4)$ is an element of $k(I) \in (\mathbb{Z}/(2\mathbb{Z}))^3 \times \mathbb{Z}$, and it determines which chart $\Omega_I = \Omega_k$ on $\mathbb{T}^3 \times \mathbb{R}$ contains the space-time support of b_I. There is also a function

$$(x(I), t(I)) : (\mathbb{Z}/(2\mathbb{Z}))^3 \times \mathbb{Z} \to \mathbb{T}^3 \times \mathbb{R},$$

which will describe the "center" of Ω_I. If we let e^i, $i = 1, 2, 3$ be the standard generators of the lattice $\mathbb{Z}^3$, then we set

$$(x(I), t(I)) = \left(\frac{1}{2} \sum_{i=1}^{3} \kappa_i e^i, \tau k_4\right).$$

Rather than view $\mathcal{I}$ as simply an index set, it is useful to think of $\mathcal{I}$ as the vertices of a graph, where we connect any two indices I if the supports

41

of V_I are able to overlap. So let us also define the set of neighbors of I, including I itself, to be the set

$$\mathcal{N}(I) = \mathcal{N}(\kappa, k_4, f) = \bigcup_{k' \in \mathcal{N}(k)} \{k'\} \times F$$

where the $2^3 \times 3$ neighboring discrete positions $\mathcal{N}(k)$ are defined by

$$\mathcal{N}(\kappa, k_4) = \{(\kappa, k_4) + u_1 e^1 + u_2 e^2 + u_3 e^3 + u_4 e^4 \mid u_i = -1, 0, 1\}.$$

Each vertex I in the above graph has $|\mathcal{N}(I)| = 2^3 \times 3 \times 12$ neighbors.

Whenever some object with an index, say Φ_I or Ω_I, only depends on only part of I, say $k = (\kappa, k_4)$, we will abuse notation and write both $\Phi_I = \Phi_k$ or $\Omega_I = \Omega_k$ to refer to the same object. We will also introduce the notation

$$\mathcal{I}(k) = k \times F$$

and

$$\mathcal{I}(k_4) = (\mathbb{Z}/(2\mathbb{Z}))^3 \times \{k_4\} \times F$$

to abbreviate important families of the indices.

The Cutoff Functions and Partitions of Unity The amplitude is equipped with two cutoff functions:

$$b_I^l = \eta((t - t(I))/\tau)\psi_I(x, t)b_{I,1}^l = \eta((t - \tau k_4)/\tau)\psi_I(x, t)b_{I,1}^l.$$

The function $\eta(t)$ is a smooth, non-negative bump function with compact support in $C_c^\infty(|t| < 5/6)$, such that

$$\sum_{k \in \mathbb{Z}} \eta^2(t - k) = 1,$$

and it is rescaled and reoriented to construct many of the other cutoff functions we use. To construct such an η, first take a smooth bump function $\tilde{\eta}$ equal to 1 for $|t| < 2/3$, and supported in $|t| < 5/6$, and set

$$\eta(t) = \frac{\tilde{\eta}(t)}{\sqrt{\sum_{k \in \mathbb{Z}} \tilde{\eta}^2(t - k)}},$$

which is smooth because the denominator is bounded away from 0. We use a partition of unity based on squares because this will allow us to patch together local solutions of

$$\sum_I |b_{I,1}|^2 (\delta^{jl} - \frac{\partial^j \xi_I \partial^l \xi_I}{|\nabla \xi_I|^2}) = G^{jl}.$$

which is homogeneous of degree two in the unknown absolute values $|b_{I,1}|$.

For each index k_4, the cutoff functions $\psi_I = \psi_{(\kappa,k_4)}$ will also be part of a similar partition of unity. We start by periodizing η to construct a partition of unity $\bar{\psi}_\kappa(x)$, $\kappa \in (\mathbb{Z}/2\mathbb{Z})^3$ of the torus, where each $\bar{\psi}_\kappa(x)$ is supported on a single chart of the torus

$$\bar{\psi}_\kappa(x_1 e^1 + x_2 e^2 + x_3 e^3) = \eta_{2\mathbb{Z}}(2x_1 - \kappa_1)\eta_{2\mathbb{Z}}(2x_2 - \kappa_2)\eta_{2\mathbb{Z}}(2x_3 - \kappa_3)$$
$$x_i \in \mathbb{R}$$
$$\eta_{2\mathbb{Z}}(y) = \sum_{k \in \mathbb{Z}} \eta(y - 2k). \tag{96}$$

Observe that $\bar{\psi}_\kappa(x)$ is integer-periodic in each x_i and is well-defined as a function of $\kappa_i \subset \mathbb{Z}/2\mathbb{Z}$. Also note that $\bar{\psi}_\kappa$ is supported in a single $\frac{5}{6} \times \frac{5}{6} \times \frac{5}{6}$ box on which our phase functions can be safely defined without critical points.

Now define $\psi_I = \psi_{\kappa,k_4}$ to be the unique solution to the initial value problem

$$(\partial_t + v_\epsilon^j \partial_j)\psi_k = 0 \tag{97}$$
$$\psi_{\kappa,k_4}(x, t(I)) = \bar{\psi}_{\kappa(I)}(x). \tag{98}$$

This way, at the initial time $t = t(I) = \tau k_4$, we have

$$\sum_{\kappa \in (\mathbb{Z}/(2\mathbb{Z}))^3} \psi_{(\kappa,k_4)}^2(x,t) = \sum_{\kappa \in (\mathbb{Z}/(2\mathbb{Z}))^3} \bar{\psi}_\kappa^2(x) = 1$$

and this identity will propagate for time $t \in \mathbb{R}$ by the transport equation.

The Phase Functions For an index $I = (\kappa, k_4, f)$, the phase function ξ_I will only be defined around the support of $\psi_{\bar{k}}(x,t)$, allowing it to exist without critical points. It is convenient to specify the domain of $\Omega_I(t)$, which is transported by the flow of v_ϵ. Set $\Omega_0 = (-7/8, 7/8) \times (-7/8, 7/8) \times (-7/8, 7/8)$, and define $\Phi_I : \Omega_0 \times \mathbb{R} \to \mathbb{T}^3 \times \mathbb{R}$ to be the unique solution to the ODE

$$\frac{d\Phi_I^j}{dt} = v_\epsilon^j(\Phi_I(x,t)) \tag{99}$$
$$\Phi_I(x, t(I)) = x + x(I) \bmod \mathbb{Z}^3. \tag{100}$$

From the well-known theorems on transport equations and ODE, the solution Φ_I to this equation is unique, it exists for all time because $\mathbb{T}^3$ is compact, and it remains a bijection onto its image $\Omega_I(t)$ because it can be inverted by reversing the flow. The fact that Φ_I is defined for all time is unimportant for us since we will be restricting to a short time interval, but observe that Ω_I covers the support of ψ_k, moreover $\psi_k(\Phi_I(x,t)) = \bar{\psi}_0(x)$.

43

We can now define the phase functions as solutions to the transport equation

$$(\partial_t + v_\epsilon^j \partial_j)\xi_I = 0 \quad \text{on } \Omega_I(t) \tag{101}$$

$$\xi_I(t(I), x) = \hat{\xi}_I(x) \quad \text{on } \Omega_I(t(I)). \tag{102}$$

Our study of the High-high frequency term motivates us to require that the initial phases $\hat{\xi}_I$ have the following properties:

- $\hat{\xi}_I$ is a linear function

- $|\nabla \hat{\xi}_I| = 1$ for all I

- There exists a $c > 0$ such that $|\nabla \hat{\xi}_I + \nabla \hat{\xi}_J| \geq c > 0$ for all neighboring indices $J \neq \bar{I}$

- For each $k = (\kappa, k_4)$, the tensors

$$\left\{ \delta^{jl} - \frac{\partial^l \hat{\xi}_{(k,f)} \partial^j \hat{\xi}_{(k,f)}}{|\nabla \hat{\xi}_{(k,f)}|^2} \mid f \in \mathbb{F} \right\}$$

form a basis for the space $\mathcal{S}$ of symmetric, $(2, 0)$ tensors at every point on Ω_k.

Once these requirements are satisfied initially, the latter two will remain true for a short time, and the first two will only hold approximately.

For the chart $k = (\kappa, k_4) = 0$, it suffices to pick phases $\{\hat{\xi}_{(0,f)} \mid f \in F\}$ with icosahedral symmetry

$$\hat{\xi}_{(0,f)}(\Phi_0(x, 0)) = < f, x > . \tag{103}$$

All the above properties can be checked for this choice, the first two being immediate and the latter two we will prove shortly.

For the other coordinate charts, we need to make sure that every phase function is separated in angle from all of its neighboring phase functions so that we still have the bound $|\nabla \xi_I + \nabla \xi_J| \geq c > 0$ for all $I \neq \bar{J}$, with a possible smaller constant c. A simple way to achieve these requirements while ensuring uniformity in the construction is to construct the initial data by rotating the dodecahedron. For this purpose, the following fact, which is proven in the Appendix, is helpful.

Lemma 7.1. *There exists a collection of 2^4 rotations O_m, indexed by $m \in (\mathbb{Z}/(2\mathbb{Z}))^4$, and a positive number $c > 0$, with the property that*

$$|f \circ O_m + f' \circ O_{m'}| \geq c \quad f, f' \in F, m, m' \in (\mathbb{Z}/(2\mathbb{Z}))^4$$

holds unless $f' = -f$ and $m' = m$. Without loss of generality, one can choose $O_0 = Id$.

With this fact in hand, we can then use the above rotations to assign initial values to all the other phase functions

$$\hat{\xi}_{(k,f)}(\Phi_I(x, t(I))) = < f \circ O_k, x > \quad k = k(I). \tag{104}$$

7.3.4   Localizing the Stress Equation

Recall that we intend to solve the equation

$$\sum_I |b_I|^2 \left(\delta^{jl} - \frac{\partial^j \xi_I \partial^l \xi_I}{|\nabla \xi_I|^2} \right) = G^{jl},$$

where the right-hand side

$$G^{jl} = e\frac{\delta^{jl}}{n} - \mathring{R}^{jl}_\epsilon \tag{105}$$

could be a fairly arbitrary, symmetric, positive $(2,0)$ tensor in $\mathcal{S}_+$.

By representing the amplitude as

$$b_I^l = \eta((t - t(I))/\tau)\psi_I(x,t)b_{I,1}^l = \eta_{k_4}\psi_k b_{I,1}^l,$$

we can use the partition of unity property of the cutoff functions to form a global solution V by gluing together the locally defined solutions to the linear systems

$$\sum_{I \in k \times F} |b_{I,1}|^2 (\delta^{jl} - \frac{\partial^j \zeta_I \partial^l \zeta_I}{|\nabla \zeta_I|^2}) = G^{jl} \quad \text{on } \Omega_k. \tag{106}$$

Indeed, once we have solved the local equation, Equation (106), for all $k = (\kappa, k_4)$, then we can check that the stress equation, Equation (93), is satisfied using the following calculation:

$$\sum_I |b_I|^2 (\delta^{jl} - \frac{\partial^j \xi_I \partial^l \xi_I}{|\nabla \xi_I|^2}) = \sum_I \eta_{k_4}^2 \psi_I^2 |b_{I,1}|^2 (\delta^{jl} - \frac{\partial^j \xi_I \partial^l \xi_I}{|\nabla \xi_I|^2})$$

$$= \sum_{k_4} \eta_{k_4}^2 \left(\sum_{I \in \mathcal{I}(k_4)} \psi_{(\kappa,k_4)}^2 |b_{I,1}|^2 (\delta^{jl} - \frac{\partial^j \xi_I \partial^l \xi_I}{|\nabla \xi_I|^2}) \right)$$

$$= \sum_{k_4} \eta_{k_4}^2 \left(\sum_{\kappa \in (\mathbb{Z}/(2\mathbb{Z}))^3} \psi_{(\kappa,k_4)}^2 \left(\sum_{I \in (\kappa,k_4) \times F} |b_{I,1}|^2 (\delta^{jl} - \frac{\partial^j \xi_I \partial^l \xi_I}{|\nabla \xi_I|^2}) \right) \right)$$

$$= \sum_{k_4} \eta_{k_4}^2 \left(\sum_{\kappa \in (\mathbb{Z}/(2\mathbb{Z}))^3} \psi_{(\kappa,k_4)}^2 \right) G^{jl}$$

$$= \sum_{k_4} \eta_{k_4}^2 G^{jl} = G^{jl}$$

Thus it suffices to solve the local equation, Equation (106), on each chart Ω_k of $\mathbb{R} \times \mathbb{T}^3$.

7.3.5 Solving the Quadratic Equation

Now observe that each term on the left-hand side of (106) is repeated twice because $b_{\bar{I}} = -b_I$ and $\nabla \xi_{\bar{I}} = -\nabla \xi_I$. To take this repetition into account, let us divide by 2, and rewrite the system (106) as

$$\sum_{I \in k \times \mathbb{F}} |b_{I,1}|^2 (\delta^{jl} - \frac{\partial^j \xi_I \partial^l \xi_I}{|\nabla \xi_I|^2}) = \frac{1}{2} G^{jl} \quad \text{on } \Omega_k \qquad (107)$$

where $\mathbb{F}$ denotes the set of faces of the projective dodecahedron introduced in Section (7.3.3). In this form, the six unknown coefficients $|b_{I,1}|^2$, $I \in k \times \mathbb{F}$, of the linear system will be uniquely determined.

If the tensor G^{jl} were completely arbitrary, we could not guarantee that the coefficients obtained by solving the linear equation would all be positive, but because of our choice of P_0, the trace part of the right-hand side dominates the trace free part, and we can rewrite the right-hand side in the form

$$\frac{G^{jl}}{2} = \frac{1}{2}(e\frac{\delta^{jl}}{n} - \mathring{R}_\epsilon^{jl}) \qquad (108)$$

$$= e(\frac{\delta^{jl}}{2n} + \varepsilon^{jl}) \qquad (109)$$

and ε^{jl} satisfies the bound

$$\varepsilon(R_\epsilon)^{jl} = -\frac{\mathring{R}_\epsilon^{jl}}{2e} \qquad (110)$$

$$|\varepsilon| \le \frac{100}{K} \qquad (111)$$

where K is the constant in the inequality (95).

With the expression (109) in mind, we write the amplitude in the form

$$b_{I,1}^l = e^{1/2} b_{I,2}^l.$$

With this normalization, we can expect the absolute value of $b_{I,2}$ to be of size ≈ 1 in order of magnitude.

Although we would like the factor $b_{I,2}$ to possess as much smoothness as possible, the pointwise constraint that $< \nabla \xi_I, b_{I,2} >= 0$ forces $b_{I,2}$ to develop some degree of irregularity. That is, even though $b_{I,2}$ will be qualitatively smooth, its derivatives will grow as its corresponding phase function is transported. Based on this observation, we first choose a vector field $\mathring{b}_I$ with size ≈ 1 which satisfies $< \nabla \xi_I, \mathring{b}_I >= 0$, and then write

$$b_{I,1}^l = e^{1/2} b_{I,2}^l = e^{1/2} \gamma_I \mathring{b}_I^l = \rho_I \mathring{b}_I^l \qquad (112)$$

where the coefficients γ_I and $\rho_I = e^{1/2} \gamma_I$ are determined by solving the linear system (107).

Constructing a Section of $\langle \nabla \xi \rangle^{\perp}$ and the Gauge Freedom We now wish to choose a vector field $\mathring{b}_I$ which satisfies $< \nabla \xi_I, \mathring{b}_I >= 0$ pointwise. We first remark that there is a huge amount of freedom in making such a choice. One can think of $\langle \nabla \xi_I \rangle^{\perp}$ as a bundle of oriented planes on which the linear map $\mathbf{J}_I = \frac{\nabla \xi_I}{|\nabla \xi_I|} \times$ acts like a complex structure, as it satisfies the equation $\mathbf{J}_I^2 = -1$ on $\langle \nabla \xi \rangle^{\perp}$. Then given any section $\mathring{b}_I$ of $\langle \nabla \xi_I \rangle^{\perp}$ (that is, any vector field satisfying $< \nabla \xi_I, \mathring{b}_I >= 0$ pointwise), and any function $\theta(t, x)$, one can obtain another section $\mathring{b}_I'$ by rotating at each point $\mathring{b}_I' = e^{\mathbf{J}_I \theta} \mathring{b}_I$ within $\langle \nabla \xi_I \rangle^{\perp}$ by an oriented angle θ. Furthermore, because the transormation $e^{\mathbf{J}_I \theta}$ preserves absolute values, any solution to the pointwise quadratic Equation (107) remains a solution to (107). This type of freedom can be regarded as a "gauge freedom," and the group of operators

$$\{e^{\mathbf{J}_I \theta}\}$$

would then be considered the "gauge group." In fact, the oscillations we produce while choosing λ to be very large are just specific examples of this gauge freedom with the function $\theta(t, x) = \lambda \xi_I(t, x)$.

Here, we do not want to use the gauge group to produce oscillations, but rather we want a specific section $\mathring{b}_I$ of $\langle \nabla \xi_I \rangle^{\perp}$ that is as smooth as possible. To be compatible with the transport of phases, we find it convenient to construct $\mathring{b}_I$ with an orthogonal projection

$$\mathring{b}_I^l = \partial^l \xi_{\sigma I} - \frac{(\nabla \xi_{\sigma I} \cdot \nabla \xi_I)}{|\nabla \xi_I|^2} \partial^l \xi_I \tag{113}$$

$$= \mathcal{P}_I^{\perp} (\nabla \xi_{\sigma I})^l. \tag{114}$$

The phase function of index $\sigma I = (k, \sigma f)$ inhabits the same coordinate patch as that of $I = (k, f)$, but it points in a different direction, so that we can guarantee that, at times sufficiently close to $t(I)$, we have

$$|\mathring{b}_I| \geq c \tag{115}$$

satisfied for some constant $c > 0$, uniformly on $\mathbb{T}^3$. In future formulas, we will express the absolute value of $\mathring{b}_I$ as

$$|\mathring{b}_I|^2 = \frac{|\nabla \xi_I \wedge \nabla \xi_{\sigma I}|^2}{|\nabla \xi_I|^2} \tag{116}$$

because $|\nabla \xi_I|^2 |\mathring{b}_I|^2 = |\nabla \xi_I|^2 |\nabla \xi_{\sigma I}|^2 - (\nabla \xi_I \cdot \nabla \xi_{\sigma I})^2$ is actually the square norm of $\nabla \xi_I \wedge \nabla \xi_{\sigma I}$ in $\Lambda^2(\mathbb{R}^3)$.

One way to choose the index σI is to let σ act as an element of order 3 in the icosahedral group. That is, let σ act on a dodecahedron as a rotation by an angle $2\pi/3$ around an axis that goes through two opposite vertices of the dodecahedron. It is easy to see that a rotation defined in this way satisfies $\sigma^3 = 1$.[12] With this choice of σ, we can then be assured that $\sigma f \neq f$ and $\sigma f \neq -f$, since the eigenvalue 1 is only obtained on the axis of rotation $\langle f + \sigma f + \sigma^2 f \rangle$, and on its orthogonal complement σ satisfies $\sigma^2 + \sigma + 1 = 0$, which has neither 1 nor -1 as a root. With this choice, we can guarantee the property (115) holds initially at time $t = t(I)$, and this property will remain true for a short time with a possibly smaller c.

Making the above choice ensures that $v_{\bar{I}} = \bar{v}_I$ and is ultimately why we chose to prescribe the imaginary part b_I in the first place rather than the real part a_I. Namely, since $\nabla \xi_{\bar{I}} = -\nabla \xi_I$, we have that $b_{\bar{I}} = -b_I$, and $a_{\bar{I}} = -\frac{(\nabla \xi_{\bar{I}})}{|\nabla \xi_{\bar{I}}|} \times b_{\bar{I}} = a_I$.

The Stress Equation Relative to a Frame   The correction V has now been written down completely except for the choice of several parameters, but we are not ready to proceed with the construction until it is clear how to estimate the derivatives of the correction. In particular, Equation (107) involves taking a square root to solve for the unknown coefficients, so we must establish positive lower bounds on the solutions of the linear equation in order to proceed. Using the expansion (112), we can rewrite (107) as

$$\sum_{I \in k \times \mathbb{F}} \rho_I^2 |\mathring{b}_I|^2 (\delta^{jl} - \frac{\partial^j \xi_I \partial^l \xi_I}{|\nabla \xi_I|^2}) = \frac{G^{jl}}{2}$$

which can also be written

$$\sum_{I \in k \times \mathbb{F}} \rho_I^2 \frac{|\nabla \xi_I \wedge \nabla \xi_{\sigma I}|^2}{|\nabla \xi_I|^2} (\delta^{jl} - \frac{\partial^j \xi_I \partial^l \xi_I}{|\nabla \xi_I|^2}) = \frac{G^{jl}}{2} \qquad (117)$$

using the identity (116).

To get more explicit control of ρ_I, it helps to express this tensorial equation relative to a basis. The most natural basis available in this regard is the set of squares $\nabla \xi_J \otimes \nabla \xi_J$, because they are compatible with the underlying transport. In this spirit, we apply both sides of Equation (117) to the square $\partial_j \xi_J \partial_l \xi_J$ with $J \in k \times \mathbb{F}$, to obtain a 6×6 system of linear equations in matrix form

$$\sum_{I \in k \times \mathbb{F}} A(\nabla \xi)_J^I \rho_I^2 = \frac{G^{jl} \partial_j \xi_J \partial_l \xi_J}{2} \qquad J \in k \times \mathbb{F}. \qquad (118)$$

Here the matrix $A(\nabla \xi)_J^I = A(\nabla \xi, \Lambda^2(\nabla \xi))_J^I$ is

$$A(\nabla \xi)_J^I = \frac{|\nabla \xi_I \wedge \nabla \xi_{\sigma I}|^2}{|\nabla \xi_I|^2} \cdot \frac{|\nabla \xi_I \wedge \nabla \xi_J|^2}{|\nabla \xi_I|^2} \qquad I, J \in k \times \mathbb{F}. \qquad (119)$$

---

[12] The icosahedral group can be identified with $A_5 \times (\mathbb{Z}/(2\mathbb{Z}))$ by considering its action on the compound of 5 cubes (or the compound of 5 tetrahedra) inscribed in the dodecahedron. The group generated by σ is therefore a 3 Sylow subgroup of the icosahedral group, which is unique up to conjugacy by Sylow's theorem, so in fact all elements of order three act as rotations in this way.

We need to prove that this linear equation can be solved, and also that the coefficients that arise are manifestly positive. At the initial time $t = t(I)$, we have the following lemma.

Lemma 7.2. *The matrix $A_J^I = A(\nabla \hat{\xi})_J^I : \mathbb{R}^\mathbb{F} \to \mathbb{R}^\mathbb{F}$ at the initial time $t(I)$ does not depend on k and is invertible.*

It is clear by the construction of the data $\hat{\xi}_I$ by rotations and the formula (119) that the initial matrix $A_J^I = A(\nabla \hat{\xi})_J^I$ is independent of k. One can check invertibility by hand using the coefficients of the matrix $A(\nabla \hat{\xi})_J^I$; in fact, one can show that all the entries in this matrix have the same value, except for the diagonal entries, which are all 0. Here we outline a slightly slower proof based on an analysis of the symmetries of the dodecahedron. We include the details in the Appendix.

Given that the numbers $\frac{|\nabla \xi_I \wedge \nabla \xi_{\sigma I}|^2}{|\nabla \xi_I|^2}$ are not zero, the matrix A_J^I is invertible if and only if both of the following statements are true.

Lemma 7.3.

1. *The tensors $\{\partial^j \xi_I \partial^l \xi_I = f(I)^j f(I)^l \mid I \in \mathbb{F}\}$ form a basis for $\mathcal{S}$*

2. *The tensors $\{\delta^{jl} - \frac{\partial^j \xi_I \partial^l \xi_I}{|\nabla \xi_I|^2} = \delta^{jl} - f^j f^l \mid I \in \mathbb{F}\}$ form a basis for $\mathcal{S}$*

These conditions are equivalent to the invertibility of A_J^I because the matrix coefficients of A_J^I are computed by taking inner products of the latter set of vectors with the former.

It is possible to reduce Claim (2) of Lemma (7.3) to Claim (1) of Lemma (7.3) by using the following formula to write down an invertible transformation taking one basis to the other:

Lemma 7.4.

$$\delta^{jl} = \frac{1}{2} \sum_{f \in \mathbb{F}} f^j f^l. \tag{120}$$

In fact, this form is, up to a constant, the unique bilinear form in $\mathcal{S}$ invariant under the action of the icosahedral group. This uniqueness property and Formula (120) both follow easily if we have already proven Claim (1) of Lemma (7.3). Indeed, if G^{jl} is any form invariant under the icosahedral group, $G^{jl} f_j f_l = C$ for just one $f \in \mathbb{F}$, then by rotating we have

$$G^{jl} f_j f_l = C \delta^{jl} f_j f_l$$

for all $f \in \mathbb{F}$. Then we can determine $C = \frac{G^{jl} \delta_{jl}}{3}$, which gives the identity (120) when we take G^{jl} to be the form on the right-hand side.

In our approach, we will use identity (120), which encodes the basic geometric properties of the dodecahedron, as a first step to prove the linear

49

independence (1). This identity is certainly very well-known, but for the benefit of the reader and for completeness of the argument, we include a proof in the Appendix, where Lemma (7.3) is also proven.

Remark about Exterior Powers. It is interesting that the second exterior power appears in the formulas here for a few reasons. First, note that the second exterior power can be implicated in explaining why the approximation $V^j V^l + P_0 \delta^{jl} + R^{jl} \approx 0$ can only be achieved in a weak topology. Namely, it is not possible to approximate an arbitrary tensor field G^{jl} by rank one tensors $V^j V^l$ in a strong topology because the form $\Lambda^2 G : \Lambda^2(\mathbb{R}^3)^* \times \Lambda^2(\mathbb{R}^3)^*$, which acts according to the formula

$$\Lambda^2 G(u_1 \wedge u_2, w_1 \wedge w_2) = \det \begin{bmatrix} G(u_1, w_1) & G(u_1, w_2) \\ G(u_2, w_1) & G(u_2, w_2) \end{bmatrix},$$

cannot be approximated by $\Lambda^2 V \otimes V = 0$, implying that almost everywhere convergence of $V^j V^l \to G^{jl}$ is not possible. On the other hand, because the function $G \mapsto \Lambda^2 G$ is nonlinear, it fails to be continuous with respect to weak convergence, and a weak approximation $V^j V^l \approx G^{jl}$ is possible when G^{jl} is non-negative definite.

The second exterior power also plays a role in our ability to control the High-High term with Beltrami flows. There, the key identity is

$$\epsilon^{abc} \epsilon_{cfg} = \delta_f^a \delta_g^b - \delta_g^a \delta_f^b$$

which provides two different formulas for the inner product on $\Lambda^2(\mathbb{R}^3)$.

Finally, the fact that a weak limit of solutions to Euler may fail to be an Euler flow is made possible by the fact that a sequence of rank one tensors $v^j v^l$ (which are characterized among non-negative, symmetric tensors by the equation $\Lambda^2 G = 0$) may have a weak limit which fails to be rank one. Thus, the success of our convex integration scheme, which implies that weak limits of Euler flows fail to be solutions, is tied in several ways to the second exterior power.

7.3.6 The Renormalized Stress Equation in Scalar Form

From the above discussion, we have established that it is possible to solve the linear system

$$A_J^I x_I = y_J \qquad I, J \in \mathbb{F} \tag{121}$$

when the matrix $A_J^I = A(\nabla \xi)_J^I = A(\nabla \hat{\xi})_J^I : \mathbb{R}^{\mathbb{F}} \to \mathbb{R}^{\mathbb{F}}$ is given by its initial value according to (119).

In general, we have to solve the nonlinear equation

$$\sum_{I \in k \times \mathbb{F}} A(\nabla \xi)_J^I \rho_I^2 = \frac{1}{2}(e \frac{\delta^{jl}}{n} - \mathring{R}_\epsilon^{jl}) \partial_j \xi_J \partial_l \xi_J \qquad J \in k \times \mathbb{F}. \tag{122}$$

Moreover, in order to control the derivatives of ρ_I, we will also have to differentiate Equation (122), which implies that we must obtain lower bounds on the coefficients ρ_I solving (122).

To see that (122) admits positive solutions ρ_I^2, we first rescale the unknowns $\rho_I = e^{1/2}\gamma_I$ and the right-hand side $G^{jl} = e(\frac{\delta^{jl}}{2n} + \varepsilon(R_\epsilon)^{jl})$, to renormalize Equation (122) into

$$\sum_{I \in k \times \mathbb{F}} A(\nabla\xi)_J^I \gamma_I^2 = (\frac{\delta^{jl}}{2n} + \varepsilon(R_\epsilon)^{jl})\partial_j \xi_J \partial_l \xi_J. \qquad (123)$$

This equation is equivalent to the tensorial form

$$\sum_{I \in k \times \mathbb{F}} \gamma_I^2 \frac{|\nabla\xi_I \wedge \nabla\xi_{\sigma I}|^2}{|\nabla\xi_I|^2}(\delta^{jl} - \frac{\partial^j \xi_I \partial^l \xi_I}{|\nabla\xi_I|^2}) = \frac{\delta^{jl}}{2n} + \varepsilon(R_\epsilon)^{jl}. \qquad (124)$$

In order to control solutions to (124), we view the equation as a perturbation of the system

$$\sum_{I \in k \times \mathbb{F}} \tilde{\gamma}_I^2 \frac{|\nabla\hat{\xi}_I \wedge \nabla\hat{\xi}_{\sigma I}|^2}{|\nabla\hat{\xi}_I|^2}(\delta^{jl} - \frac{\partial^j \hat{\xi}_I \partial^l \hat{\xi}_I}{|\nabla\hat{\xi}_I|^2}) = \frac{\delta^{jl}}{2n}. \qquad (125)$$

The system (125) admits positive solutions $\tilde{\gamma}_I^2$. In fact, as we demonstrate in the Appendix, the formula (120) together with a symmetry argument implies the identities

$$|\nabla\hat{\xi}_{(k,f)}|^2 = |f|^2 = 1 \qquad (126)$$

$$|\nabla\hat{\xi}_{(k,f)} \wedge \nabla\hat{\xi}_{(k,\sigma f)}|^2 = |f|^2|\sigma f|^2 - (f \cdot \sigma f)^2 = \frac{4}{5}. \qquad (127)$$

Using these identities, we see that Equation (125), along with its equivalent scalar form

$$\sum_{I \in k \times \mathbb{F}} A(\nabla\hat{\xi})_J^I \tilde{\gamma}_I^2 = (\frac{\delta^{jl}}{2n})\partial_j \hat{\xi}_J \partial_l \hat{\xi}_J$$

$$= \frac{|\nabla\hat{\xi}_J|^2}{6} = 1/6 \qquad (128)$$

is solved by any set of $\tilde{\gamma}_I$ satisfying

$$\tilde{\gamma}_I^2 = \frac{5}{2^5 \cdot 3}. \qquad (129)$$

In particular, each component of the solution $\tilde{\gamma}_I = \sqrt{\frac{5}{2^5 \cdot 3}}$ is nonzero.

We hope to treat Equation (123) as a perturbation of Equation (128). The first step in this direction is to show that Equation (128) can be perturbed to one which still can be solved, and whose solutions remain bounded from 0. This preparatory step is accomplished by the following lemma:

Lemma 7.5 (The Implicit Function Lemma). *There is a positive $c > 0$ and $a < 1$ such that whenever*

$$||A_J^I - A(\nabla\hat\xi)_J^I|| \le c \tag{130}$$

$$||y_J - \frac{|\nabla\hat\xi_J|^2}{6}|| \le c \tag{131}$$

the equation

$$\sum_I A_J^I \gamma_I^2 = y_J \tag{132}$$

has a unique solution in the ball $|\gamma_I - \tilde\gamma_I| \le a|\tilde\gamma_I|$.

The solution γ_I is represented by a function

$$\gamma_I = \gamma_{f(I)}(A, y) \tag{133}$$

that depends smoothly on A_J^I and y_J in the domain (130). The function $\gamma_{f(I)}$ is one of 6 functions $\{\gamma_f : f \in \mathbb{F}\}$, and is determined by the direction coordinate $f(I) \in \mathbb{F}$.

Furthermore, the matrix $B_J^I = 2A_J^I \gamma_I$ obtained by linearizing the equation is invertible in this domain with uniform bounds

$$||A^{-1}|| \le C \tag{134}$$

$$||(\gamma_I)^{-1}|| \le C. \tag{135}$$

Remark. It is important to observe that the constant $c > 0$ is completely independent of the location k of the phase directions. This independence follows from the fact that the matrix $A(\nabla\hat\xi)_J^I$ and the right-hand side $\frac{|\hat\xi_J|^2}{6}$ of Equation (128) that we perturb are independent of the location index.

Proof. By writing Equation (132) in the form

$$F(\gamma, A, y) = 0 \tag{136}$$

$$F(\gamma, A, y)_J \equiv \sum_I A_J^I \gamma_I^2 - y_J \tag{137}$$

we can deduce the lemma as an immediate consequence of the implicit function theorem once we verify that the matrix

$$\frac{\partial F_J}{\partial \gamma_I}(\gamma, A, y) = (A_J^I)(2\gamma_I) \tag{138}$$

$$\frac{\partial F_J}{\partial \gamma_I} : \mathbb{R}^{\mathbb{F}} \to \mathbb{R}^{\mathbb{F}} \tag{139}$$

is invertible at the point $(\gamma_I, A_J^I, y_J) = (\tilde\gamma_I, A(\nabla\hat\xi)_J^I, |\nabla\hat\xi_J|^2/6)$.

To prove that the above matrix is one-to-one, it suffices to show that its null space is zero. If h_I is any element in the null space at this point, then using the fact that $A(\nabla\hat{\xi})^I_J$ is invertible and that all the $\tilde{\gamma}_I$ are nonzero, we deduce that

$$\frac{\partial F_J}{\partial \gamma_I} h_I = 2A(\nabla\hat{\xi})^I_J \tilde{\gamma}_I h_I = 0 \qquad \forall J \in \mathbb{F} \tag{140}$$

$$\Rightarrow \tilde{\gamma}_I h_I = 0 \qquad \forall I \in \mathbb{F} \tag{141}$$

$$\Rightarrow h_I = 0 \qquad \forall I \in \mathbb{F}. \tag{142}$$

The implicit function theorem then guarantees the existence of c and a as in the lemma, because it also guarantees that the matrix $\frac{\partial F_J}{\partial \gamma_I}$ remains invertible in the domain (130). $\qquad\qquad\square$

Lemma (7.5) applies to Equation (123) with

$$A^I_J = A(\nabla\xi)^I_J \tag{143}$$

$$y_J - (\frac{\delta^{jl}}{2n} + \varepsilon (R_\epsilon)^{jl})\partial_j\xi_J\partial_l\xi_J \tag{144}$$

$$= \frac{|\nabla\hat{\xi}|^2}{6} + \frac{1}{6}(|\nabla\xi|^2 - |\nabla\hat{\xi}|^2) + \varepsilon^{jl}\partial_j\xi_J\partial_l\xi_J \tag{145}$$

provided that K is sufficiently large and that τ is sufficiently small.

At this point, we can choose K once and for all. Let c be the positive constant guaranteed by Lemma (7.5), and we pick

$$K = 7000c^{-1} \tag{146}$$

so that

$$|\varepsilon| \leq \frac{100}{K} \leq \frac{c}{70}. \tag{147}$$

Having chosen K, we can prove the following theorem which restricts how small τ must be chosen:

Proposition 7.1. *There exist constants $c > 0$ and $a > 0$ with $a < 1$ such that Equation (123) admits a unique solution $\gamma_I \in \mathbb{R}^{\mathbb{F}}$ in a neighborhood $\{|\gamma_I - \tilde{\gamma}_I| \leq a|\tilde{\gamma}_I|\}$ provided that*

$$\big|\, |\nabla\xi_I|^2 - |\nabla\hat{\xi}_I|^2\, \big| \leq c \qquad \forall I \in F \tag{148}$$

$$\big|\, |\nabla(\xi_I + \xi_J)|^2 - |\nabla(\hat{\xi}_I + \hat{\xi}_J)|^2\, \big| \leq c \qquad \forall I, J \in F. \tag{149}$$

The solution $\gamma_I(\nabla\xi_I, \varepsilon)$ depends smoothly on $\nabla\xi_I \in \mathbb{F}$ and ε for $\nabla\xi_I$ in the domain (148) and ε in the domain (147).

Proof. In the statement of the proposition, we have chosen to assume control over the deviation of the gradient energies

$$\big|\,|\nabla(\xi_I + \xi_J)|^2 - |\nabla(\hat{\xi}_I + \hat{\xi}_J)|^2\,\big|$$

rather than the matrices $A(\nabla\xi) - A(\nabla\hat{\xi})$ because the entries of the matrix $A(\nabla\xi)^I_J$ depend only on the inner products $\nabla\xi_I \cdot \nabla\xi_J$, which themselves can be calculated from the polarization identity

$$
\begin{aligned}
|\nabla(\xi_I + \xi_J)|^2 - |\nabla(\xi_I - \xi_J)|^2 &= |\nabla(\xi_I + \xi_J)|^2 - |\nabla(\xi_I + \xi_{\bar{J}})|^2 \qquad (150)\\
&= 2\nabla\xi_I \cdot \nabla\xi_J. \qquad\qquad\qquad\qquad (151)
\end{aligned}
$$

The proposition then immediately follows from Lemma (7.5) once we observe the bound

$$
\begin{aligned}
\left| y_J - \frac{|\nabla\hat{\xi}_J|^2}{6} \right| &\leq \frac{\big|\,|\nabla\xi_J|^2 - |\nabla\hat{\xi}_J|^2\,\big|}{6} + |\varepsilon||\nabla\xi_J|^2 \\
&\leq \frac{\big|\,|\nabla\xi_J|^2 - |\nabla\hat{\xi}_J|^2\,\big|}{6} + \frac{c}{70}|\nabla\xi_J|^2 \\
&\leq \frac{c}{70}|\nabla\hat{\xi}_J|^2 + (\frac{1}{6} + \frac{c}{70})|\,|\nabla\xi_J|^2 - |\nabla\hat{\xi}_J|^2\,| \\
&= \frac{c}{70} + A|\,|\nabla\xi_J|^2 - |\nabla\hat{\xi}_J|^2\,|.
\end{aligned}
$$

$\square$

Thanks to the above proposition, we are allowed to differentiate Equation (118) in order to obtain upper bounds on the derivatives of ρ_I.

In carrying out the construction, we always require that τ is chosen small enough so that the Proposition (7.1) applies. Besides the requirement of Proposition (7.1), we also require that the gradients $|\nabla(\xi_I + \xi_J)| \geq c > 0$ remain bounded from 0 by a constant.

7.3.7   Summary

To summarize the results of the preceding sections, we have established the following:

Proposition 7.2. *There exist absolute constants $c > 0$ and $K > 0$ such that as long as*

$$\|\nabla\xi_I - \nabla\hat{\xi}_I\|_{C^0} \leq c \qquad \forall I \in k \times F \qquad (152)$$

and

$$e(t) \geq K|R_\epsilon(t, x)| \qquad\qquad (153)$$

pointwise, then the local form (106) of the stress equation, Equation (90), admits a unique solution of the form

$$v_I = a_I + i b_I \tag{154}$$

$$I = (k, f) = (k_1, k_2, k_3, k_4, f) \tag{155}$$

$$b_I^l = \eta \left(\frac{t - t(I)}{\tau} \right) \psi_k \gamma_I \mathcal{P}_I^\perp (\nabla \xi_{\sigma I})^l \tag{156}$$

$$a_I = -\frac{(\nabla \xi_I)}{|\nabla \xi_I|} \times b_I \tag{157}$$

where

$$\mathcal{P}_I^\perp (\nabla \xi_{\sigma I})^l = \partial^l \xi_{\sigma I} - \frac{(\nabla \xi_{\sigma I} \cdot \nabla \xi_I)}{|\nabla \xi_I|^2} \partial^l \xi_I \tag{158}$$

is the orthogonal projection of the phase gradient $\nabla \xi_{\sigma I}$ into $\langle \nabla \xi_I \rangle^\perp$ and the coefficients

$$\gamma_I = \gamma_I (\nabla \xi_k, \varepsilon) \tag{159}$$

$$= \gamma_f (\nabla \xi_k, \varepsilon) \tag{160}$$

depend on the phase gradients in region k and the tensor

$$\varepsilon^{jl} = -\frac{\mathring{R}_\epsilon}{e(t)}. \tag{161}$$

The smooth function γ_f in (160) is one of 6 smooth functions indexed by the direction coordinate $f(I) \in \mathbb{F}$ of $I = (k, f)$, which is independent of the location coordinate $k \in (\mathbb{Z}/2\mathbb{Z})^3 \times \mathbb{Z}$.

For reference, the functions η and ψ_k are elements of partitions of unity which are explained in Section (7.3.4). The phase functions ξ_I are only defined on a portion Ω_I of $\mathbb{R} \times \mathbb{T}^3$ defined in (7.3.3) on which they can live without critical points, and they satisfy the transport equation

$$(\partial_t + v_\epsilon^j \partial_j) \xi_I = 0$$

$$\xi_I(t(I), x) = \hat{\xi}_I(x)$$

for initial data $\hat{\xi}_I$ at time $t(I) = \tau k_4$ specified in Section (7.3.3).

Part IV
Obtaining Solutions from the Construction

8 Constructing Continuous Solutions

We will now show how the preceding construction, combined with a few estimates from Part (V), can be used to prove Lemma (5.1).

Proof of Lemma (5.1). Let $\epsilon > 0$ and (v, p, R) be given, and let $e(t)$ be the function specified by the lemma.

We define K to be the constant chosen in Line (146).

We describe the argument as a sequence of 6 steps.

8.1 Step 1: Mollifying the Velocity

The error term generated by mollifying v has the form

$$Q_v^{jl} = (v^j - v_\epsilon^j)V^l + V^j(v^l - v_\epsilon^l) \tag{162}$$

$$= 2[(v - v_\epsilon)V]^{jl} \tag{163}$$

where V is a **locally finite** sum

$$V = \sum_I V_I \tag{164}$$

$$V_I = e^{i\lambda\xi_I}(v_I + \frac{\nabla \times w_I}{\lambda}) \tag{165}$$

whose cardinality at each point (t, x) is bounded by an absolute constant. Namely, at each time t there are at most two generations k_4 for which V_I are nonzero, and for each generation k_4, there are at most 8 families (k_1, k_2, k_3, k_4), each having at most $|F| = 2 \cdot 6$ phase functions, which are nonzero.

We choose λ at the end of the argument in order to shrink the term

$$Q_{v,(ii)}^{jl} = \sum_I 2e^{i\lambda\xi_I} \frac{[(v - v_\epsilon)\nabla \times w_I]^{jl}}{\lambda} \tag{166}$$

since the bounds for $\nabla \times w_I$ depend on v_ϵ.

For v_ϵ, we choose a space-time mollification $v_\epsilon = \eta_{\epsilon_t, x} * v$ close enough so that we can guarantee that for the main term

$$Q_{v,(i)} = \sum_I 2[(v - v_\epsilon)(e^{i\lambda\xi_I}v_I)] \tag{167}$$

we have

$$\|Q_{v,(i)}\|_{C^0} \leq \|(v - v_\epsilon)\|_{C^0} \|\sum_I v_I\|_{C^0} \qquad (168)$$

$$\lhd \frac{\epsilon}{40}. \qquad (169)$$

Logically speaking, the term v_I is not defined until v_ϵ has been constructed, however, we can give a priori bounds for its size since we know that it will have the form

$$v_I = a_I + ib_I,$$

where

$$b_I^l = \eta\left(\frac{t - t(I)}{\tau}\right) \psi_k(t, x) e^{1/2}(t) \gamma_I(\nabla \xi_k, \frac{R_\epsilon}{e}) P_I^\perp (\nabla \xi_{\sigma I})^l \qquad (170)$$

and $a_I = -\frac{(\nabla \xi_I) \times}{|\nabla \xi_I|} b_I$ is obtained by a $\pi/2$ rotation of b_I in $\langle \nabla \xi_I \rangle^\perp$.

Until v_ϵ is constructed, no bounds for derivatives of v_I can be stated, but it is clear that, once they are well defined, each v_I **will** be bounded in absolute value by

$$\|v_I\|_{C^0} \lhd 2000 K^{-1/2} e_R^{1/2}, \qquad (171)$$

as we have assumed an upper bound (20) for $e^{1/2}(t)$.

Therefore the locally finite sum

$$\|\sum_I e^{i\lambda \xi_I} v_I\|_{C^0} \lhd A e_R^{1/2} \qquad (172)$$

will also be bounded, a priori, since the cardinality of the terms which are nonzero at any given (t, x) is also bounded.

This estimate can only be confirmed when the lifespan parameter τ is chosen, but anticipating the above estimate, we can choose the mollifier $\eta_{\epsilon_{t,x}}$ for v so that (169) will be a consequence of (171) once (171) has been proven.

8.2 Step 2: Mollifying the Stress

Let $R_\epsilon = \eta_{\epsilon_{t,x}} * R$ be a mollification of R in the variables (t, x) so that

- supp $R_\epsilon \subseteq I' \times \mathbb{T}^3$ where I' is an interval on which $e(t) \geq K e_R$

- $\|R - R_\epsilon\|_{C^0} \leq \frac{\epsilon}{100}$.

8.3 Step 3: Choosing the Lifespan

Since we have assumed that the velocity v is uniform bounded, we can set

$$\theta^{-1} \geq \epsilon_v^{-1} \|v\|_{C^0}$$

to be an upper bound[13] for $\|\nabla v_\epsilon\|_{C^0}$. Then set

$$\tau = b\theta \qquad (173)$$

for some parameter b. According to the bounds for transport equations proven in Section (17), we have that

$$\|\nabla \xi_I - \nabla \hat{\xi}_I\|_{C^0} \leq Ab \qquad (174)$$

for some constant A. We will therefore choose b small enough, so that, first of all, the conditions in Line (148) are guaranteed. Our second goal for choosing a small parameter τ is to ensure that the first term in the parametrix for the High-High term is controlled.

Namely, we should be able to make the C^0 norm of

$$\tilde{Q}_{H,(1)}^{jl} = \sum_{J \neq \bar{I}} e^{i\lambda(\xi_I + \xi_J)} i^{-1} q_{H,IJ,(1)}^{jl} \qquad (175)$$

smaller than

$$\|\tilde{Q}_{H,(1)}^{jl}\|_{C^0} \trianglelefteq \frac{\epsilon}{500}. \qquad (176)$$

Here, $q_{H,IJ,(1)}^{jl}$ is the solution to the linear equation

$$\partial_j \xi_I q_{H,IJ,(1)}^{jl} = \left(v_J \times ([|\nabla \xi_I| - 1]v_I)^l + v_I \times ([|\nabla \xi_J| - 1]v_J)^l\right) \qquad (177)$$

described in Section (6.2). Although the v_I and v_J are not defined until τ is chosen, as long as τ satisfies the conditions (148), which we have already guaranteed, we know that

$$\|v_J \times ([|\nabla \xi_I| - 1]v_I)\|_{C^0} \leq Ae_R \||\nabla \xi_I| - 1\|_{C^0} \qquad (178)$$

which will be bounded by another constant

$$\|v_J \times ([|\nabla \xi_I| - 1]v_I)\|_{C^0} \leq A'e_R b \qquad (179)$$

according to the bounds in Section (17).

Therefore, it is possible to choose b such that (176) will be guaranteed by the construction. Note that τ only depends on the given v and e_R.

---

[13]If we remove the assumption of uniform boundedness, it is still possible to estimate the derivatives of v_ϵ in terms of only the modulus of continuity of v. This task can be accomplished by using the expression

$$\partial_i \eta_\epsilon * v^l = \int v^l(x+h)\partial_i \eta_\epsilon(h)dh$$

$$= \int (v^l(x+h) - v^l(x))\partial_i \eta_\epsilon(h)dh.$$

Later on during the construction of Hölder continuous solutions, the choice of τ will turn out to be completely independent of $\|v\|_{C^0}$.

8.4 Step 4: Bounds for the New Stress

All the remaining terms in the new stress R_1 are $O(1/\lambda)$ in C^0 once the above parameters have been chosen.

For example, the term (166) and the Stress term

$$Q_S^{jl} = \sum_I V_I^j \bar{V}_I^l + P_0 \delta^{jl} + R_\epsilon^{jl} \tag{180}$$

$$= \sum_I \frac{(\nabla \times w_I)^j \bar{v}_I^l}{\lambda} + \sum_I \frac{v_I^j (\nabla \times \bar{w}_I)^l}{\lambda} \tag{181}$$

$$+ \sum_I \frac{(\nabla \times w_I)^j (\nabla \times \bar{w}_I)^l}{\lambda^2} \tag{182}$$

is $O(\lambda^{-1})$ in C^0.

The remaining terms in the stress require us to solve the divergence equation. They include, for example, the High-Low term

$$\partial_j Q_L^{jl} = \sum_I e^{i\lambda \xi_I} \tilde{v}_I^j \partial_j v_\epsilon^l \tag{183}$$

$$= e^{i\lambda \xi_I} u_L^l. \tag{184}$$

To control these terms, we require estimates on the phase gradients and their spatial derivatives in order to implement the parametrix of Section (6.2). In Section (17) we show that, for τ as chosen above, each derivative of $\nabla \xi$ costs, essentially, a factor of $C \epsilon_v^{-1}$. Therefore the vector field u_L, its derivatives, and the derivatives of the phase functions can be bounded uniformly in terms of ϵ_v and v. We have also guaranteed bounds for $\||\nabla \xi_I|^{-1}\|_{C^0}$ by the choice of τ.

Applying Lemma (6.1) with sufficiently large λ, there exists a solution of size

$$\|Q_L\|_{C^0} \leq C_{v,R} \lambda^{-1}. \tag{185}$$

The Transport term

$$\partial_j Q_T^{jl} = \sum_I e^{i\lambda \xi_I} [(\partial_t + v_\epsilon^j \partial_j) \tilde{v}_I^l] \tag{186}$$

likewise has a solution of size

$$\|Q_T^{jl}\|_{C^0} \leq C_{v_\epsilon, R_\epsilon} \lambda^{-1} \tag{187}$$

by Lemma (6.1).

Finally, what remains of the High-High term after the first iteration of the parametrix must solve the equation

$$\partial_j Q_{H,(2)}^{jl} = \sum_{J \neq \bar{I}} -e^{i\lambda(\xi_I + \xi_J)} i^{-1} \partial_j q_{H,IJ,(1)}^{jl} \tag{188}$$

where $q^{jl}_{H,IJ,(1)}$ is defined as in (177). This equation also has a solution of size

$$\|Q_H\|_{C^0} \le C_{v_\epsilon, R_\epsilon} \lambda^{-1} \tag{189}$$

by Lemma (6.1), where here we apply our bounds on third derivatives $\nabla^3 \xi$ from Section (17).

We demand that λ be at least large enough so that the sum of the bounds for these solutions is smaller than $\frac{\epsilon}{3}$ in C^0. The last task that remains is to show that the energy increment can be controlled precisely.

8.5 Step 5: Bounds for the Corrections

Recall that

$$V_I = e^{i\lambda \xi_I}(v_I + \frac{\nabla \times w_I}{\lambda}) \tag{190}$$

$$= e^{i\lambda \xi_I} \tilde{v}_I. \tag{191}$$

We have already observed that

$$\|v_I\|_{C^0} \le A e_R^{1/2} \tag{192}$$

so to prove that $\|V_I\|_{C^0}$ also satisfies this bound (with a slightly larger constant), it suffices to choose λ large enough.

Similarly, the correction to the pressure takes the form

$$P = P_0 + \sum_{J \ne \bar{I}} P_{I,J} \tag{193}$$

$$P_0 = -\frac{e(t)}{3} + \frac{R^{jl}_\epsilon \delta_{jl}}{3} \tag{194}$$

$$P_{I,J} = -\frac{V_I \cdot V_J}{2} \tag{195}$$

$$= -\frac{1}{2} e^{i\lambda(\xi_I + \xi_J)} \tilde{v}_I \cdot \tilde{v}_J \tag{196}$$

so that when λ is at least as large as above, we also have

$$\|P\|_{C^0} \le C e_R. \tag{197}$$

Thus we have established the bounds for the corrections.

8.6 Step 6: Control of the Energy Increment

Consider the energy increment

$$\int [|v_1|^2(t,x) - |v|^2(t,x)]dx = \int |V|^2(t,x)dx + 2\int V \cdot v dx. \tag{198}$$

For the second term, we expect that because V is highly oscillatory, it can be made orthogonal to v. To prove this fact, let v_S be any smooth vector field such that

$$Ce_R^{1/2}\left\|\int |v - v_S|(t,x)dx\right\|_{C_t^0} \le \frac{\epsilon}{3} \tag{199}$$

where $Ce_R^{1/2}$ is the upper bound for $\|V\|_{C^0}$ established after Equation (190).

Now we estimate the remaining term

$$\int V \cdot v_S dx = \int \nabla \times W \cdot v_S dx \tag{200}$$

$$= \int W \cdot \nabla \times v_S dx \tag{201}$$

$$= O(\lambda^{-1}). \tag{202}$$

It remains to bound

$$\left\|\int |V|^2(t,x)dx - \int e(t)dx\right\|_{C_t^0} \le \frac{\epsilon}{3}. \tag{203}$$

In order to achieve this approximation, we compute

$$\int |V|^2(t,x)dx = \int V^j V^l \delta_{jl} dx \tag{204}$$

$$= \int \sum_{I,J} V_I^j V_J^l \delta_{jl} dx \tag{205}$$

$$= \sum_I \int V_I^j \bar{V}_I^l \delta_{jl} dx + \sum_{J \neq \bar{I}} \int V_I \cdot V_J dx \tag{206}$$

$$= \sum_I \int \tilde{v}_I^j \bar{\tilde{v}}_I^l \delta_{jl} dx + \sum_{J \neq \bar{I}} \int e^{i\lambda(\xi_I + \xi_J)} \tilde{v}_I \cdot \tilde{v}_J dx \tag{207}$$

$$= \int \left(\sum_I v_I^j \bar{v}_I^l\right) \delta_{jl} dx + \sum_{J \neq \bar{I}} \int e^{i\lambda(\xi_I + \xi_J)} \tilde{v}_I \cdot \tilde{v}_J dx \tag{208}$$

$$+ \sum_I \int \frac{(\nabla \times w_I)}{\lambda} \cdot \bar{v}_I dx + \sum_I \int \frac{(\nabla \times \bar{w}_I)}{\lambda} \cdot v_I dx$$

$$+ \sum_I \int \frac{(\nabla \times w_I)}{\lambda} \cdot \frac{(\nabla \times w_I)}{\lambda} dx \tag{209}$$

$$= \int e(t)dx + \sum_{J \neq \bar{I}} \int e^{i\lambda(\xi_I + \xi_J)} \tilde{v}_I \cdot \tilde{v}_J dx + O(\lambda^{-1}) \tag{210}$$

where the $O(\lambda^{-1})$ bound on the error term is uniform in time.

To bound the error term, we integrate by parts

$$\int e^{i\lambda(\xi_I+\xi_J)}\tilde{v}_I \cdot \tilde{v}_J dx = \int \left[\frac{\partial^a(\xi_I+\xi_J)}{i\lambda|\nabla(\xi_I+\xi_J)|^2}\partial_a[e^{i\lambda(\xi_I+\xi_J)}] \right]\tilde{v}_I \cdot \tilde{v}_J dx \quad (211)$$

$$= \frac{-1}{i\lambda}\int e^{i\lambda(\xi_I+\xi_J)}\partial_a \left[\frac{\partial^a(\xi_I+\xi_J)}{|\nabla(\xi_I+\xi_J)|^2}\tilde{v}_I \cdot \tilde{v}_J \right] dx. \quad (212)$$

Taking λ large concludes the proof of Lemma (5.1). $\square$

9   Frequency and Energy Levels

Now that we have constructed continuous solutions, let us discuss how to measure the Hölder regularity of the solutions we construct when the scheme is executed more carefully. For this aspect of the convex integration scheme, our biggest contribution is a notion of **frequency and energy levels**. This notion is meant to accurately record the bounds which apply to the (v, p, R) coming from the previous stage of the construction. This notion is based on the paper [CDLS12b], but differs from that paper in that here there is a gap between the estimates for ∇v and ∇R that must be recorded, and the definition below also records higher derivatives.

Here is an example of a candidate definition for frequency and energy levels, although it is not the one that will be used in our book.

Definition 9.1 (Practice Frequency Energy Levels). Let $L \geq 1$ be a fixed integer. Let $\Xi \geq 2$, and let e_v and e_R be positive numbers with $e_R \leq e_v$. Let (v, p, R) be a solution to the Euler-Reynolds system. We say that the practice frequency and energy levels of (v, p, R) are below (Ξ, e_v, e_R) (to order L in $C^0 = C^0_{t,x}(\mathbb{T}^3 \times \mathbb{R})$) if the following estimates hold.

$$\|\nabla^k v\|_{C^0} \leq \Xi^k e_v^{1/2} \qquad k = 1, \ldots, L \qquad (213)$$

$$\|\nabla^k p\|_{C^0} \leq \Xi^k e_v \qquad k = 1, \ldots, L \qquad (214)$$

$$\|\nabla^k R\|_{C^0} \leq \Xi^k e_R \qquad k = 0, \ldots, L \qquad (215)$$

Here ∇ refers only to derivatives in the spatial variables.

The idea of these bounds is that at stage k, $e_{v,(k)}$ is the size $(e_{R,(k-1)})$ of the stress $R_{(k-1)}$ coming from the previous stage, and that $\Xi_{(k)}$ is essentially the parameter $\lambda_{(k-1)}$ chosen in the previous stage. Then the bounds (213) are consistent with the heuristic representation

$$V_{(k-1)} \approx e^{i\Xi_{(k)}x}|R_{(k-1)}|^{1/2}$$

since $V_{(k-1)}$ gives the dominant contribution to the derivatives of $v_{(k)} = v_{(k-1)} + V_{(k-1)}$.

Based on a definition of frequency and energy levels such as Definition (9.1) we can summarize the effect of one iteration of the convex integration procedure in a single lemma, which has roughly the following form.

Sample Lemma 9.1. For every solution (v, p, R) with frequency and energy levels below

$$(\Xi, e_v, e_R)$$

and every

$$N \geq \left(\frac{e_v}{e_R}\right) \tag{216}$$

there exists a solution (v_1, p_1, R_1) to the Euler-Reynolds equations with new frequency and energy levels below

$$(\Xi', e_v', e_R') = (N\Xi, e_R, e_R') \tag{217}$$

such that $v_1 = v + V$, $p_1 = p + P$, and

$$\|V\|_{C^0} \leq Ce_R^{1/2} \tag{218}$$

$$\|\nabla V\|_{C^0} \leq CN\Xi e_R^{1/2} \tag{219}$$

$$\|P\|_{C^0} \leq Ce_R \tag{220}$$

$$\|\nabla P\|_{C^0} \leq CN\Xi e_R. \tag{221}$$

The idea of the above outline is that

$$N\Xi \approx \lambda$$

will be essentially the chosen value for λ. The bounds (218), (219) for the new energy and frequency levels then follow from the heuristic representation

$$V \approx e^{i(N\Xi)x}|R|^{1/2}.$$

The efficiency of the convex integration process is measured by how small one can make e_R' if one is only allowed to grow the frequency by a factor N.

The condition (216) ensures, for example, that the largest term in

$$\|\nabla v_1\|_{C^0} = \|\nabla v\|_{C^0} + \|\nabla V\|_{C^0}$$
$$\leq C(\Xi e_v^{1/2} + N\Xi e_R^{1/2})$$
$$\leq CN\Xi e_R^{1/2}$$

is the one coming from the high frequency V, and similarly for P

$$\|\nabla p_1\|_{C^0} = \|\nabla p\|_{C^0} + \|\nabla P\|_{C^0}$$
$$\leq C(\Xi e_v + N\Xi e_R)$$
$$\leq CN\Xi e_R.$$

Of course, to be useful, the definition (9.1) should be modified to contain information about time derivatives as well, and at the moment it is not clear that bounds for the pressure would be relevant given the passive role played by the pressure so far.

In terms of this rough notion of frequency and energy levels, let us see what kind of rate e'_R we can expect for the reduced stress in the present problem.

First consider the High-Low Interaction term

$$\partial_j Q_L^{jl} = \sum_I e^{i\lambda\xi_I} \tilde{v}_I^j \partial_j v_\epsilon^l. \tag{222}$$

According to Definition (9.1) and the expected bound $\|\tilde{v}_I\|_{C^0} \leq A e_R^{1/2}$, the right-hand side has size

$$\|\tilde{v}_I^j \partial_j v_\epsilon^l\|_{C^0} \leq e_R^{1/2}(\Xi e_v^{1/2}). \tag{223}$$

On the other hand, the parametrix gains a factor $\lambda^{-1} \approx N^{-1}\Xi^{-1}$, so we expect a solution

$$\|Q_L\|_{C^0} \leq \frac{e_v^{1/2} e_R^{1/2}}{N}. \tag{224}$$

If we could actually prove a theorem such as Sample Lemma (9.1) with

$$e'_R = \frac{e_v^{1/2} e_R^{1/2}}{N} \tag{225}$$

then our construction could achieve solutions in the class

$$v \in C^{1/3-\epsilon} \tag{226}$$

$$p \in C^{2(1/3-\epsilon)} \tag{227}$$

which have the regularity conjectured by Onsager. To apply the lemma, we choose $N_{(k)}$ at each stage so that the energy levels decay just slightly super-exponentially with a self-similarity Ansatz

$$e_{R,(k+1)} = \frac{e_{R,(k)}^{1+\delta}}{Z} \tag{228}$$

$$Z = \text{constant}. \tag{229}$$

Letting $\delta \to 0$ causes the gaps in the frequency spectrum between consecutive stages of the construction to be small, and allows the regularity to increase up to $1/3$ if the rate (225) were to hold. This ideal case is explained further in Section (13). The above considerations mirror the analysis in the introduction to [DLS14].

However, before we can come close to the ideal rate of (225), we must significantly change the Definition (9.1). To see why a drastic change is necessary, consider the contribution to the new stress which arises from the Transport term

$$\partial_j Q_T^{jl} \approx \sum_I e^{i\lambda\xi_I}(\partial_t + v_\epsilon^j \partial_j)v_I^l. \tag{230}$$

The v_I itself has amplitude $e_R^{1/2}$ since it is constructed from solving the stress equation. Formally, we can think of $v_I \approx R^{1/2}$. The derivative in $\partial_j v_I^l$ costs a factor of Ξ, and the factor

$$v_\epsilon^j \approx 1$$

is bounded by some constant, giving the second term in (230) size $\Xi e_R^{1/2}$. Then, solving the divergence equation gains a factor of $(N\Xi)^{-1}$, leading to an expected bound for the new stress

$$\|Q_T\|_{C^0} \lesssim \frac{e_R^{1/2}}{N} \tag{231}$$

$$\Rightarrow e_R' \gtrsim \frac{e_R^{1/2}}{N} \tag{232}$$

which is a factor $e_v^{-1/2}$ worse than the ideal case (225), and which is also inconsistent with dimensional analysis.

In order to calculate the regularity achieved from applying Lemma (9.1) with a rate of (232), one still imposes the Ansatz (228), but in this case it is **not optimal** to let δ tend to 0. Doing so would result in a new frequency

$$\Xi_{(k+1)} = e_{R,(k)}^{-1/2-\delta}\Xi_{(k)} \tag{233}$$

which is much larger than the previous frequency, but would result in only a very small reduction of stress. On the other hand, letting δ be very large would be wasteful since the frequencies $\Xi_{(k)}$ would grow very quickly.

In this case, iteration of (228) at a rate of (232) leads to a regularity up to

$$v \in C^{\alpha^*-\epsilon} \tag{234}$$

$$p \in C^{2(\alpha^*-\epsilon)} \tag{235}$$

with

$$\alpha^* = \frac{\delta}{(1+\delta)(1+2\delta)}. \tag{236}$$

The optimal δ to choose is $\delta^* = \frac{1}{\sqrt{2}}$, which would lead to solutions with regularity [14] up to

$$\alpha^* = \frac{1}{3 + \sqrt{8}}. \tag{237}$$

These regularity exponents can be computed quickly using the machinery introduced in Section (11).

In order to achieve a regularity better than (237) we examine the size of the Transport term more carefully, without separating the terms in the coarse scale material derivative.

$$\frac{\bar{D}}{\partial t} = (\partial_t + v_\epsilon \cdot \nabla)$$

In fact, when a coarse scale material derivative hits a phase gradient, which obeys the equation

$$(\partial_t + v_\epsilon^a \partial_a)\partial_l \xi_I = -\partial_l v_\epsilon^a \partial_a \xi_I \tag{238}$$

the cost is

$$|\frac{\bar{D}}{\partial t}| \leq \Xi e_v^{1/2}$$

which is exactly what we desire for (225).

The trouble is that the amplitude $v_I \approx R^{1/2}$ is also influenced by the stress R. To deal with this difficulty, we must assume that the bound for the material derivative of R is **better** than the bound for the spatial derivative of R in our definition of frequency and energy levels. In order to make this assumption on the material derivative of R, we must also verify it in order to continue the iteration. To obtain these improved bounds, we introduce the following changes to the scheme:

1. We change the way in which R_ϵ is mollified so that the material derivative of the Transport term

$$\partial_j Q_T^{jl} \approx \sum_I e^{i\lambda \xi_I} (\partial_t + v_\epsilon^j \partial_j) v_I^l \tag{239}$$

   can be estimated. Namely, checking the bound for the material derivative of Q_T requires taking a *second* material derivative of R_ϵ, whereas we only assume a bound on the first material derivative of R.

2. We also change the way we eliminate the error for the parametrix by solving the divergence equation

$$\partial_j Q^{jl} = U^l \tag{240}$$

---

[14]Of course, to actually obtain this regularity would at least require a modification of the Definition (9.1) to incorporate time derivatives, and Lemma (9.1) must be stated precisely and proven.

since it is not true that every solution to the underdetermined Equation (240) will have a good bound on its material derivative. We find that the simplest way to obtain bounds on the material derivative of a solution to (240) is to solve (240) by constructing an appropriate transport equation.

3. Furthermore, we cannot perform a standard mollification of v in the time variable because bounds on time derivatives are not as good as bounds on material derivatives. In fact, we only mollify $v \to v_\epsilon$ in space, and the regularity of v_ϵ in time follows from the Euler-Reynolds equations themselves.

These modifications are presented in the remainder of the book, which we begin by properly formalizing the template Lemma (9.1).

10   The Main Iteration Lemma

10.1   Frequency Energy Levels for the Euler-Reynolds Equations

The Main Lemma in the book, which is responsible for the proof of Theorem (0.1), relies on the following definition.

Definition 10.1. Let $L \geq 1$ be a fixed integer. Let $\Xi \geq 2$, and let e_v and e_R be positive numbers with $e_R \leq e_v$. Let (v, p, R) be a solution to the Euler-Reynolds system. We say that (v, p, R) have **frequency and energy levels** below (Ξ, e_v, e_R) (to order L in $C^0 - C^0_{t,x}(\mathbb{T}^3 \times \mathbb{R})$) if the following estimates hold.

$$\begin{align}
\|\nabla^k v\|_{C^0} &\leq \Xi^k e_v^{1/2} & k &= 1, \ldots, L & (241)\\
\|\nabla^k p\|_{C^0} &\leq \Xi^k e_v & k &= 1, \ldots, L & (242)\\
\|\nabla^k R\|_{C^0} &\leq \Xi^k e_R & k &= 0, \ldots, L & (243)\\
\|\nabla^k (\partial_t + v \cdot \nabla) R\|_{C^0} &\leq \Xi^{k+1} e_v^{1/2} e_R & k &= 0, \ldots, L-1 & (244)
\end{align}$$

Here ∇ refers only to derivatives in the spatial variables.

We first remark that these bounds are all consistent with the symmetries of the Euler equations. The scaling symmetry is reflected by dimensional analysis. Informally, if X denotes a spatial unit (e.g., "meters") and T denotes a unit of time (e.g., "seconds"), then the dimensions of each term involved in the estimates are $e_R, e_V \sim p, R \sim X^2/T^2$, $\Xi \sim \nabla \sim X^{-1}$, $\partial_t \sim T^{-1}$ and $v \sim X/T$.

We have also assumed **no** bound on $\|v\|_{C^0}$ or on $\|\partial_t R\|_{C^0}$. These assumptions are consistent with the Galilean invariance of the Euler equations and the Euler-Reynolds equations, which plays a special role underlying what follows. Namely, if (v, p, R) solve the Euler-Reynolds equations, then a new solution to Euler-Reynolds with the same frequency energy levels can be obtained by taking

$$\hat{v}^j(t, x) = C^j + v^j(t, x - tC) \tag{245}$$

$$\hat{p}(t, x) = p(t, x - tC) \tag{246}$$

$$\hat{R}^{jl}(t, x) = R^{jl}(t, x - tC) \tag{247}$$

for any $C \in \mathbb{R}^3$. This transformation corresponds to observing the fluid from a new frame of reference with relative velocity C. Furthermore, the aspects of the construction we introduce to take the special role of the material derivative into account (namely, the mollifications of v and R and the new method used to solve the divergence equation) are also well-behaved under this group of transformations.

Also note that we have not assumed a bound on $\|p\|_{C^0}$ and that only the pressure gradients are assumed to have bounds, which is consistent with the fact that the pressure is only determined from the velocity up to an arbitrary function of time. It is interesting to remark that one can include bounds on the material derivative of the pressure gradient such as

$$\|(\partial_t + v \cdot \nabla)\nabla p\|_{C^0} \leq \Xi^2 e_v$$

in the definition of frequency energy levels. However, these bounds are not necessary for the results in the present book.

10.2   Statement of the Main Lemma

With this definition in hand, we can state the Main Lemma. Note that we require the order $L \geq 2$, as we would not be able to prove a lemma in the form of the Main Lemma below if we assume control over only one derivative; this requirement is discussed in Section (15). The requirement of at least two derivatives appears to be natural in view of the calculations of Section (13.1), and may be necessary for controlling one of the error terms if the method could be pushed to regularity $1/3$, as is noted in Section (25.1).

Lemma 10.1 (The Main Lemma). *Suppose $L \geq 2$ and that $\eta > 0$. Let K be the constant in Line (146). There is a constant C depending only on η and L such that the following holds:*

Let (v, p, R) be any solution of the Euler-Reynolds system whose frequency and energy levels are below (Ξ, e_v, e_R) to order L in C^0 and let I be a union of nonempty time intervals (possibly unbounded) such that

$$\text{supp } R \subseteq I \times \mathbb{T}^3. \tag{248}$$

Define the time-scale $\theta = \Xi^{-1} e_v^{-1/2}$, and let

$$e(t) : \mathbb{R} \to \mathbb{R}_{\geq 0}$$

be any non-negative function satisfying the lower bound

$$e(t) \geq K e_R \qquad \text{for all } t \in I \pm \theta \tag{249}$$

and whose square root satisfies the estimates

$$||\frac{d^r e^{1/2}}{dt^r}||_{C^0} \leq 1000 (\Xi e_v^{1/2})^r (K e_R)^{1/2} \qquad r = 0, 1, 2. \tag{250}$$

Now let N be any positive number obeying the bounds

$$N \geq \Xi^\eta \tag{251}$$

$$N \geq \left(\frac{e_v}{e_R}\right)^{3/2} \tag{252}$$

and define the dimensionless parameter $\mathbf{b} = \left(\frac{e_v^{1/2}}{e_R^{1/2} N}\right)^{1/2}$.

Then there exists a solution (v_1, p_1, R_1) of the Euler-Reynolds system of the form $v_1 = v + V$, $p_1 = p + P$ whose frequency and energy levels are below:

$$(\Xi', c_v', c_R') = (CN\Xi, c_R, \frac{e_v^{1/4} e_R^{3/4}}{N^{1/2}})$$

$$= (CN\Xi, e_R, \left(\frac{e_v^{1/2}}{e_R^{1/2} N}\right)^{1/2} e_R) \tag{253}$$

$$= (CN\Xi, e_R, \mathbf{b}^{-1} \frac{e_v^{1/2} e_R^{1/2}}{N})$$

to order L in C^0, and whose stress R_1 is supported in

$$\text{supp } R_1 \subseteq (\text{ supp } e \pm \theta) \times \mathbb{T}^3. \tag{254}$$

The correction $V = v_1 - v$ is of the form $V = \nabla \times W$ and can be guaranteed to obey the bounds

$$||V||_{C^0} \leq C e_R^{1/2} \tag{255}$$

$$||\nabla V||_{C^0} \leq CN\Xi e_R^{1/2} \tag{256}$$

$$||(\partial_t + v^j \partial_j) V||_{C^0} \leq C\mathbf{b}^{-1} \Xi e_v^{1/2} e_R^{1/2} \tag{257}$$

$$||W||_{C^0} \leq C\Xi^{-1} N^{-1} e_R^{1/2} \tag{258}$$

$$||\nabla W||_{C^0} \leq C e_R^{1/2} \tag{259}$$

$$||(\partial_t + v^j \partial_j) W||_{C^0} \leq C\mathbf{b}^{-1} N^{-1} e_v^{1/2} e_R^{1/2} \tag{260}$$

and its energy can be prescribed up to errors bounded by

$$\left\| \int_{\mathbb{T}^3} |V|^2 dx - \int_{\mathbb{T}^3} e(t) dx \right\|_{C^0} \leq C \frac{e_R}{N} \tag{261}$$

$$\left\| \frac{d}{dt} \left(\int_{\mathbb{T}^3} |V|^2 dx - \int_{\mathbb{T}^3} e(t) dx \right) \right\|_{C^0} \leq C \mathbf{b}^{-1} \frac{\Xi e_v^{1/2} e_R}{N}. \tag{262}$$

The correction to the pressure $P = p_1 - p_0$ satisfies the estimates

$$\|P\|_{C^0} \leq C e_R \tag{263}$$

$$\|\nabla P\|_{C^0} \leq C N \Xi e_R \tag{264}$$

$$\|(\partial_t + v \cdot \nabla) P\|_{C^0} \leq C \mathbf{b}^{-1} \Xi e_v^{1/2} e_R \tag{265}$$

Finally, the supports of both V and P are contained in $\operatorname{supp} e \times \mathbb{T}^3$.

Remark about Dimensional Analysis and (251). Although it is not consistent with inequality (251), the parameter N should be regarded as a dimensionless parameter. When the lemma is applied, the parameter η will tend to 0, and the constant C in the lemma will diverge to infinity. The assumption (251) will be used to bound the number of terms taken in the parametrix before the approximate solution is sufficiently accurate.

A Conjecture and a Remark about Energy Regularity. Having stated the Main Lemma, we propose a "conjecture" which is probably false as stated, but which summarizes what the proof of Lemma (10.1) would prove if the High-High Interference terms could be sufficiently well-controlled.

Conjecture 10.1 (The Ideal Case). *The conclusion of the Main Lemma (10.1) holds with the factor $\mathbf{b}^{-1}$ replaced by 1 in all time derivative and material derivative estimates, and with*

$$(\Xi', e_v', e_R') = (C N \Xi, e_R, \frac{e_v^{1/2} e_R^{1/2}}{N}).$$

As we discuss in Section (13), the above conjecture could be used to construct solutions of regularity up to $1/3$, and would therefore imply Onsager's conjecture. In Section (19), we discuss the term which prevents us from proving the above conjecture.

We also remark that although the solutions produced by Lemma (10.1) go up to the Hölder exponent $1/5$, the energy function

$$E(t) = \int \frac{|v|^2}{2}(t, x) dx$$

turns out to possess better regularity in time, as it necessarily belongs to the class $C^{1/2-\epsilon}(\mathbb{R})$ from the way it is constructed. The reason for this higher regularity will be discussed in Section (13).

We now show that Lemma (10.1) contains enough quantitative information regarding the construction to prove the main theorem. For convenience, we state the form of the main theorem which is proven directly by iterating Lemma (10.1).

Theorem 10.1. *Let $\delta > 0$ and set*

$$\alpha^* = \frac{1+\delta}{5+9\delta+4\delta^2}. \tag{266}$$

Then there exists a nonzero solution to the Euler equations with compact support in time such that for all $\alpha < \alpha^*$

$$\|v\|_{C^{0,\alpha}_{t,x}} + \|p\|_{C^{0,2\alpha}_{t,x}} < \infty. \tag{267}$$

Furthermore, the energy of v can be made arbitrarily large, and the time interval containing the support of (v, p) can be made arbitrarily small.

11 Main Lemma Implies the Main Theorem

The proof of Theorem (10.1) proceeds by inductively applying the Main Lemma (10.1) in order to construct a sequence of solutions $(v_{(k)}, p_{(k)}, R_{(k)})$ of the Euler-Reynolds system such that $R_{(k)}$ converges to 0 uniformly and such that $v_{(k)}$ converges in $C^{0,\alpha}_{t,x}$ for all $\alpha < \alpha^*$. The parameters $\eta > 0$ and L are fixed at the beginning of the argument to depend only on δ. In fact we can simply take $L = 2$. At each stage (k) of the induction, we choose an energy function $e_{(k)}(t)$ and a parameter $N_{(k)}$ whose choice determines the growth of the frequency parameter $\Xi_{(k)}$ and the decay of the energy level $e_{R,(k)}$.

In fact, we will iterate the Main Lemma in such a way that

$$e_{v,(k+1)} = e_{R,(k)} \tag{268}$$

$$e_{R,(k+1)} = \frac{e_{R,(k)}^{(1+\delta)}}{Z} \tag{269}$$

where Z is a constant chosen at the beginning of the iteration in order to make sure that $e_{R,(1)} < e_{R,(0)}$, and so that other inequalities are satisfied. The self-similarity Ansatz (269) leads to super-exponential decay of the energy levels $e_{v,(k)}, e_{R,(k)}$ and super-exponential growth of the frequencies $\Xi_{(k)}$.

In order to be able to achieve this Ansatz, we must apply the lemma with

$$N_{(k)} = Z^2 \left(\frac{e_v}{e_R}\right)^{1/2}_{(k)} e_{R,(k)}^{-2\delta}, \tag{270}$$

which will imply a frequency growth of

$$\Xi_{(k+1)} = C_{\eta,L} Z^2 \left(\frac{e_v}{e_R}\right)^{1/2}_{(k)} e^{-2\delta}_{R,(k)} \Xi_{(k)}. \tag{271}$$

The exponent η (which determines how many terms must be taken in the parametrix expansion for the elliptic equation in the main argument, and therefore comes into the constant in the above equality) is also fixed at the beginning of the iteration.

11.1   The Base Case

We begin the proof by establishing a base case lemma, which will ensure that the previous choice of $N_{(k)}$ will always satisfy the admissibility conditions (251),(252).

Lemma 11.1 (Base Case). *Suppose that $\delta > 0$ and $L \geq 2$. Then there exists an $\eta_0 > 0$ such that for all positive numbers $e_{R,(0)} > 0, C > 0, \Xi_{(0)} \geq 2$ and $\eta > 0$ with $\eta \leq \eta_0$, for every nonempty interval $I_{(0)}$ there exist arbitrarily large numbers Z and there exists a solution (v_0, p_0, R_0) to the Euler-Reynolds system*

$$\partial_t v^l + \partial_j(v^j v^l) + \partial^l p = \partial_j R^{jl} \tag{272}$$

$$\partial_i v^i = 0 \tag{273}$$

that has support contained in $I_{(0)} \times \mathbb{T}^3$ and that has frequency and energy levels below $(\Xi_{(0)}, e_{v,(0)}, e_{R,(0)})$ to order L in C^0, where $(\Xi_{(0)}, e_{v,(0)}, e_{R,(0)})$ and Z satisfy all of the following:

1. *For*

$$\gamma = 1/(1+\delta)$$

   we have the relationship

$$\begin{aligned} e_{v,(0)} &= Z^\gamma e^\gamma_{R,(0)} \\ e_{R,(0)} &< e_{v,(0)} \end{aligned} \tag{274}$$

   which can also be written as

$$e_{R,(0)} = \frac{e^{1+\delta}_{v,(0)}}{Z}. \tag{275}$$

2. *Condition (252) is satisfied for*

$$N_{(0)} = Z^2 \left(\frac{e_v}{e_R}\right)^{1/2}_{(0)} e^{-2\delta}_{R,(0)}. \tag{276}$$

Namely,

$$\left(\frac{e_v}{e_R}\right)_{(0)}^{3/2} \le Z^2 \left(\frac{e_v}{e_R}\right)_{(0)}^{1/2} e_{R,(0)}^{-2\delta}. \tag{277}$$

3. Condition (251) is also satisfied for $N_{(0)}$, namely

$$\Xi_{(0)}^\eta \le Z^2 \left(\frac{e_v}{e_R}\right)_{(0)}^{1/2} e_{R,(0)}^{-2\delta}. \tag{278}$$

4. We can also ensure that, for $k = 0$,

$$\left[Z^{-1}e_{R,(k)}^\delta\right]^{\delta(2+\frac{\gamma}{2})} \left[CZ^{2+\frac{\gamma}{2}} e_{R,(k)}^{-\delta(2+\frac{\gamma}{2})}\right]^\eta < 1 \tag{279}$$

and the exponent to which $e_{R,(k)}$ is raised above is positive. This inequality will be used later in the proof in order to make sure that (251) continues to be verified for the choices of $N_{(k)}$ at every later stage of the iteration.

Proof. We can take

$$(v_0, p_0, R_0) = (0, 0, 0),$$

which can be regarded as a solution to Euler-Reynolds with frequency and energy bounded below any trio $(\Xi_{(0)}, e_{v,(0)}, e_{R,(0)})$ that are admissible according to the Definition (10.1). The lemma reduces to verifying that for every $e_{R,(0)}$ and $C > 0$, there exist numbers $\Xi_{(0)}$ and Z so that for $e_{v,(0)} = Z^\gamma e_{R,(0)}^\gamma$, the trio $(\Xi_{(0)}, e_{v,(0)}, e_{R,(0)})$ satisfies all of the inequalities in the lemma simultaneously. Thus, let $e_{R,(0)}$ and $C > 0$ be given.

Inequality (277) To see that inequality (277) can be satisfied, we rewrite (277) as a goal

$$Z^{-2}\left(\frac{e_v}{e_R}\right)_{(0)} e_{R,(0)}^{2\delta} \trianglelefteq 1. \tag{280}$$

From the identity (274), we have that

$$\left(\frac{e_v}{e_R}\right)_{(0)} = Z^\gamma e_{R,(0)}^{\gamma-1} \tag{281}$$

$$= Z^\gamma e_{R,(0)}^{-\delta\gamma} \tag{282}$$

since $\gamma(1 + \delta) = 1$.

Substituting into the left-hand side of (280) gives

$$Z^{-2}\left(\frac{e_v}{e_R}\right)_{(0)} e_{R,(0)}^{2\delta} = Z^{\gamma-2} e_{R,(0)}^{-\delta(\gamma-2)} \tag{283}$$

which will be less than 1 for any Z sufficiently large, depending on $e_{R,(0)}$.

Inequality (278) In order to verify inequality (278), we rewrite it as a goal

$$Z^{-2}\left(\frac{e_v}{e_R}\right)_{(0)}^{-1/2} e_{R,(0)}^{2\delta}\Xi_{(0)}^{\eta} \trianglelefteq 1 \tag{284}$$

$$Z^{-2}\left[Z^{\gamma}e_{R,(0)}^{-\delta\gamma}\right]^{-1/2} e_{R,(0)}^{2\delta}\Xi_{(0)}^{\eta} \trianglelefteq 1 \tag{285}$$

$$Z^{-(2+\frac{\gamma}{2})}e_{R,(0)}^{\frac{(\delta\gamma)}{2}+2\delta}\Xi_{(0)}^{\eta} \leq 1. \tag{286}$$

For any given $\Xi_{(0)}$, η and $e_{R,(0)}$, inequality (286) will be satisfied if Z is chosen large enough.

Motivation for Inequality (279) Let us discuss the importance of the inequality (279).

In order to ensure that inequality (278), or the equivalent inequality (286), are satisfied for choices of the parameter $N_{(k)}$ in later stages, we need to show that

$$Z^{-(2+\frac{\gamma}{2})}e_{R,(k)}^{\frac{\delta\gamma}{2}+2\delta}\Xi_{(k)}^{\eta} \trianglelefteq 1 \tag{287}$$

for all $(k) \geq 0$. This inequality has been established for $(k) = 0$, so in order to accomplish this goal for later (k), it suffices to show that the sequence

$$x_{(k)} \equiv e_{R,(k)}^{\frac{\delta\gamma}{2}+2\delta}\Xi_{(k)}^{\eta} \tag{288}$$

$$= e_{R,(k)}^{\delta(2+\frac{\gamma}{2})}\Xi_{(k)}^{\eta} \tag{289}$$

is decreasing in (k). (Here we have applied the identity $\gamma(1+\delta) = 1$.)

From (269), (271) and (282), we can deduce the following evolution rules for the parameters

$$e_{R,(k+1)} = \frac{e_{R,(k)}^{\delta}}{Z}e_{R,(k)} \tag{290}$$

$$\Xi_{(k+1)} = CZ^2\left(\frac{e_v}{e_R}\right)_{(k)}^{1/2} e_{R,(k)}^{-2\delta}\Xi_{(k)} \tag{291}$$

$$= CZ^2\left[Z^{\gamma}e_{R,(k)}^{-\delta\gamma}\right]^{1/2} e_{R,(k)}^{-2\delta}\Xi_{(k)} \tag{292}$$

$$= CZ^{2+\frac{\gamma}{2}}e_{R,(k)}^{\frac{-\delta\gamma}{2}-2\delta}\Xi_{(k)} \tag{293}$$

implying that

$$x_{(k+1)} = \left[Z^{-1}e_{R,(k)}^{\delta}\right]^{\delta(2+\frac{\gamma}{2})}\left[CZ^{2+\frac{\gamma}{2}}e_{R,(k)}^{-\delta(2+\frac{\gamma}{2})}\right]^{\eta}x_{(k)}. \tag{294}$$

Inequality (279)   Observe that for $\eta < \eta_0$ sufficiently small, the exponent of Z in inequality (279) is strictly negative and the exponent to which $e_{R,(k)}$ is raised is strictly positive provided $\eta_0 < \frac{\delta}{2}$. Fix such an η. Then for any given C and $e_{R,(0)}$, it is possible to choose Z large enough so that (279) is satisfied.   □

11.2   The Main Lemma Implies the Main Theorem

We now begin the proof that Lemma (10.1) implies Theorem (10.1).

Proof of the Main Theorem (10.1). Let $\delta > 0$, $e_{R,(0)} > 0$ and $\Xi_{(0)} \geq 2$ be given numbers. We will construct a sequence of solutions of the Euler-Reynolds system $(v_{(k)}, p_{(k)}, R_{(k)})$ such that $R_{(k)}$ converges to 0 uniformly and such that $v_{(k)}$ converges in $C_{t,x}^{0,\alpha}$ for all

$$\alpha < \alpha^* = \frac{1+\delta}{5+9\delta+4\delta^2}.$$

We will also see that $p_{(k)}$ converges in $C_{t,x}^{0,\alpha}$ for all $\alpha < 2\alpha^*$.

Step 0:
   Choose $\eta > 0$ small enough such that the total exponent

$$\delta^2(2 + \frac{\gamma}{2}) - \eta\delta(2 + \frac{\gamma}{2})$$

to which $e_{R,(k)}$ is raised in Equation (294) is positive. For example, $\eta = \frac{\delta}{2}$ will suffice. For this choice of η and $L = 2$, let $C = C_\eta$ be the constant given by the Main Lemma, Lemma (10.1).
   To initialize the construction, apply the base case Lemma (11.1) with the parameters

$$(\delta, C_\eta, e_{R,(0)}, \Xi_{(0)})$$

to obtain a constant $Z > 0$ and a solution $(v_{(0)}, p_{(0)}, R_{(0)})$ to the Euler-Reynolds equations with frequency energy levels below $(\Xi_{(0)}, e_{v,(0)}, e_{R,(0)})$, where

$$e_{R,(0)} = \frac{e_{v,(0)}^{1+\delta}}{Z}. \tag{295}$$

We can require this solution to have support in the interval $I_{(0)} = \{0\}$, since the solution guaranteed by the lemma is identically 0.

11.2.1   Choosing the Parameters

Step k:

To continue the construction, we will iterate Lemma (10.1) to produce a sequence $(v_{(k)}, p_{(k)}, R_{(k)})$ with corresponding energy and frequency levels below $(\Xi_{(k)}, e_{v,(k)}, e_{R,(k)})$, and we impose the Ansatz

$$e_{R,(k+1)} = Z^{-1} e_{R,(k)}^{(1+\delta)}. \tag{296}$$

The inequality (274) implies that $e_{R,(1)} = Z^{-1} e_{R,(0)}^{(1+\delta)} < e_{R,(0)}$, so by induction the sequence $e_{R,(k)}$ defined recursively by (296) is a decreasing sequence.

In fact, the Ansatz (296) together with the requirement $e_{R,(1)} < e_{R,(0)}$ guarantees that the sequence $e_{R,(k)}$ decreases to 0 at a super-exponential rate, as one can see by writing the relation as

$$e_{R,(k+1)} = (Z^{-1} e_{R,(k)}^{\delta}) e_{R,(k)}. \tag{297}$$

Then $Z^{-1} e_{R,(k)}^{\delta} < 1$ for $k = 0$, so by induction even the ratio between consecutive terms

$$\frac{e_{R,(k+1)}}{e_{R,(k)}} = Z^{-1} e_{R,(k)}^{\delta}$$

decreases to 0.

Because $e_{v,(k+1)} = e_{R,(k)}$, the Ansatz (296) also implies by induction a gap between the energy levels e_v and e_R

$$e_{v,(k)} = Z^{\gamma} e_{R,(k)}^{\gamma} \quad k \geq 1 \tag{298}$$

$$\gamma(1 + \delta) = 1 \tag{299}$$

implying that the ratio between the energy levels

$$\left(\frac{e_v}{e_R}\right)_{(k)} = Z^{\gamma} e_{R,(k)}^{-\delta\gamma} \tag{300}$$

increases super-exponentially in (k).

In order to achieve this Ansatz, we must verify that the choice of

$$N_{(k)} = Z^2 \left(\frac{e_v}{e_R}\right)_{(k)}^{1/2} e_{R,(k)}^{-2\delta} \tag{301}$$

always satisfies the admissibility conditions (251) and (252).

In Line (283) in the proof of Lemma (11.1), we showed that the condition (251) is equivalent to the inequality

$$Z^{\gamma-2} e_{R,(k)}^{-\delta\gamma+2\delta} \leq 1. \tag{302}$$

For $(k) = 0$, Lemma (11.1) guarantees that this inequality is satisfied. Since the power $\delta(2 - \gamma)$ to which $e_{R,(k)}$ is raised is positive, and $e_{R,(k)}$

decreases, we see that (302) and therefore (251) are always satisfied by induction.

Similarly, inequality (252) is satisfied for $(k) = 0$ by Lemma (11.1), and the proof showed that (252) continues to be satisfied if we have that

$$\left[Z^{-1} e_{R,(k)}^{\delta} \right]^{\delta(2+\frac{\gamma}{2})} \left[C_{\eta} Z^{2+\frac{\gamma}{2}} e_{R,(k)}^{-\delta(2+\frac{\gamma}{2})} \right]^{\eta} \le 1. \tag{303}$$

According to Lemma (11.1), this inequality is satisfied for $(k) = 0$ if Z is sufficiently large. By the choice of $\eta < \delta$, the power to which $e_{R,(k)}$ is raised on the left-hand side of inequality (303) is positive, so the inequality (303) and therefore (252) continues to hold for all $(k) \ge 0$. This completes the proof that the choice of $N_{(k)}$ is always admissible.

11.2.2 Choosing the Energies

Along with these choices of $(\Xi_{(k)}, e_{v,(k)}, e_{R,(k)})$ and $N_{(k)}$, the Main Lemma also grants us the ability to approximately prescribe the energy increment $e_{(k)}(t)$ of the correction.

The energy must satisfy the estimates:

$$\left\| \frac{d^l e_{(k)}^{1/2}}{dt^l} \right\|_{C^0} \le 1000 K^{1/2} (\Xi e_{v,(k)}^{-1/2})^l e_{R,(k)}^{1/2} \qquad l \quad 0, 1, 2. \tag{304}$$

and must satisfy the lower bound

$$e_{(k)}(t) \ge K e_{R,(k)} \tag{305}$$

for t belonging to the active region $t \in I_{(k)} \pm \theta_{(k)}$, where the stress is supported $\mathrm{supp}\, R_{(k)} \subseteq I_{(k)} \times \mathbb{T}^3$.

For our construction, let $I_{(k)} = [a_{(k)}, b_{(k)}]$ be an interval such that

$$\mathrm{supp}\, R_{(k)} \subseteq I_{(k)}$$

and let $\tau_{(k)}$ be a small number which we will specify. Denote by

$$I_{(k)} \pm (\theta_{(k)} + \tau_{(k)}) = [a_{(k)} - (\theta_{(k)} + \tau_{(k)}), b_{(k)} + (\theta_{(k)} + \tau_{(k)})],$$

and let

$$\chi_{(k)}(t) = \chi_{I_{(k)} \pm (\theta_{(k)} + \tau_{(k)})}(t)$$

be the characteristic function of this time interval.

Set

$$e_{(k)}^{1/2}(t) = (20 K e_{R,(k)})^{1/2} \eta_{\tau_{(k)}} * \chi_{(k)}(t) \tag{306}$$

where

$$\eta_{\tau_{(k)}}(t) = \tau_{(k)}^{-1} \eta(\frac{t}{\tau_{(k)}}) \tag{307}$$

is a standard, smooth mollifier of integral $\int \eta_{\tau_{(k)}}(t)dt = 1$, supported in the ball $\{|t| \le |\tau_{(k)}|\}$.

Then, on the interval $I_{(k)} \pm \theta_{(k)}$ containing the support of the triple $(v_{(k)}, p_{(k)}, R_{(k)})$, we clearly have $e_{(k)}(t) = 20Ke_{R,(k)} \ge Ke_{R,(k)}$, and we also have bounds of the form

$$\|e_{(k)}^{1/2}(t)\|_{C^0} \le (20K)^{1/2}e_{R,(k)}^{1/2} \tag{308}$$

$$\|\frac{d^l e_{(k)}^{1/2}}{dt^l}\|_{C^0} \le B\tau_{(k)}^{-l}K^{1/2}e_{R,(k)}^{1/2} \qquad l = 1, 2 \tag{309}$$

Therefore $e_{(k)}(t)$ is admissible as an approximate energy function if we choose

$$\tau_{(k)} = B_0 \Xi_{(k)}^{-1} e_{v,(k)}^{-1/2} = B_0 \theta_{(k)} \tag{310}$$

where B_0 is a constant.

11.2.3 Regularity of the Velocity Field

We now check that using the parameters chosen above, the sequence of solutions $(v_{(k)}, p_{(k)}, R_{(k)})$ in the above argument does in fact converge to a solution with the properties stated in Theorem (10.1).

Continuity   The C^0 convergence of

$$v_{(K)} = \sum_{k=0}^{K} V_{(k)}$$

and follows easily from the super-exponential decay of $e_{R,(k)}$, the estimate

$$\|V_{(k)}\|_{C^0} \le Ce_{R,(k)}^{1/2}$$

and the triangle inequality

$$\sum_{k=0}^{\infty} \|V_{(k)}\|_{C^0} \le C \sum_{k=0}^{\infty} e_{R,(k)}^{1/2}. \tag{311}$$

Similarly, the C^0 convergence of

$$p_{(K)} = \sum_{k=0}^{K} P_{(k)}$$

follows from the estimate

$$\|P_{(k)}\|_{C^0} \le Ce_{R,(k)}.$$

78

Hölder Norms of the Correction To measure what kind of higher regularity is achieved by this choice, observe that the perturbation $V_{(k)} = v_{(k+1)} - v_{(k)}$ obeys the estimates

$$\|V_{(k)}\|_{C^0} \le C e_{R,(k)}^{1/2}$$

$$\|\nabla V_{(k)}\|_{C^0} \le C N_{(k)} \Xi_{(k)} e_{R,(k)}^{1/2}$$

$$\|\partial_t V_{(k)}\|_{C^0} \le \|(\partial_t + v_{(k)} \cdot \nabla) V_{(k)}\|_{C^0} + \|v_{(k)} \cdot \nabla V_{(k)}\|_{C^0}.$$

The estimate for the time derivative is no worse than the bound on the spatial derivative for the following reasons. We have already established that the $\|v_{(k)}\|_{C^0}$ are uniformly bounded, so we have a bound

$$\|v_{(k)} \cdot \nabla V_{(k)}\|_{C^0} \le C \|\nabla V_{(k)}\|_{C^0}$$

with a constant depending only on $e_{R,(0)}$ and other absolute constants from the lemmas.

Despite the loss of a factor

$$\mathbf{b}_{(k)}^{-1} = \left(\frac{N_{(k)} e_{R,(k)}^{1/2}}{e_{v,(k)}^{1/2}} \right)^{1/2} \tag{312}$$

$$= \left(Z^2 e_{R,(k)}^{2\delta} \right)^{1/2} = Z e_{R,(k)}^{\delta} \tag{313}$$

in the estimates for material derivatives in Lemma (10.1), the estimate for the material derivative is still better than the bounds on the spatial derivatives. Namely

$$\|(\partial_t + v_{(k)} \cdot \nabla) V_{(k)}\|_{C^0} \le C \mathbf{b}_{(k)}^{-1} \Xi_{(k)} e_{v,(k)}^{1/2} e_{R,(k)}^{1/2} \tag{314}$$

$$\le C Z e_{R,(k)}^{-\delta} \Xi_{(k)} e_{v,(k)}^{1/2} e_{R,(k)}^{1/2} \tag{315}$$

$$= C Z e_{v,(k)}^{1/2} \Xi_{(k)} e_{R,(k)}^{1/2-\delta} \tag{316}$$

whereas the spatial derivatives are bounded by

$$\|\nabla V_{(k)}\|_{C^0} \le C N_{(k)} \Xi_{(k)} e_{R,(k)}^{1/2} \tag{317}$$

$$\le C Z^2 \left(\frac{e_v}{e_R} \right)_{(k)}^{1/2} e_{R,(k)}^{-2\delta} \Xi_{(k)} e_{R,(k)}^{1/2} \tag{318}$$

$$= C \left(\frac{e_v}{e_R} \right)_{(k)}^{1/2} \Xi_{(k)} e_{R,(k)}^{1/2-2\delta} \tag{319}$$

which is worse than the estimate for the material derivative by a super-exponentially large factor of $e_{R,(k)}^{-1/2-\delta}$.

These observations lead to the bounds

$$\|V_{(k)}\|_{C^0} \leq C e_{R,(k)}^{1/2} \tag{320}$$

$$\|\nabla_{t,x} V_{(k)}\|_{C^0} \leq C \left(\frac{e_v}{e_R}\right)_{(k)}^{1/2} e_{R,(k)}^{-2\delta} \Xi_{(k)} e_{R,(k)}^{1/2}, \tag{321}$$

which, by interpolation, imply a bound

$$\|V_{(k)}\|_{C_{t,x}^\alpha} \leq C \left(\frac{e_v}{e_R}\right)_{(k)}^{\alpha/2} \Xi_{(k)}^\alpha e_{R,(k)}^{1/2-2\delta\alpha} \tag{322}$$

for $0 \leq \alpha \leq 1$.

Therefore, to determine whether or not the series

$$\sum_{(k)} \|V_{(k)}\|_{C_{t,x}^\alpha} < \infty \tag{323}$$

converges, it suffices to study the growth or decay of the quantity

$$E_{\alpha,(k)} \equiv \left(\frac{e_v}{e_R}\right)_{(k)}^{\alpha/2} \Xi_{(k)}^\alpha e_{R,(k)}^{1/2-2\delta\alpha}. \tag{324}$$

At this point, one way we can proceed would be to follow the method of Section (11.1) and establish an evolution law of the form

$$E_{\alpha,(k+1)} \leq C_\alpha e_{R,(k)}^{\delta f(\alpha)} E_{\alpha,(k)} \tag{325}$$

using parameter evolution laws (269), (271) and (282). Once such a law has been established, we have a trichotomy:

- If $f(\alpha)$ is positive, then $E_{\alpha,(k)}$ decays super-exponentially

- If $f(\alpha)$ is negative, then $E_{\alpha,(k)}$ grows super-exponentially

- If $f(\alpha)$ is zero, then $E_{\alpha,(k)}$ grows exponentially

The exponent $f(\alpha)$ depends linearly on α, and it turns out that there is a critical $\alpha^* \in (0,1)$ depending on δ at which $f(\alpha^*) = 0$. For all $\alpha < \alpha^*$, the series (323) converges, and the solution $v \in C_{t,x}^\alpha$. In our case, $\alpha^* = \frac{1+\delta}{5+9\delta+4\delta^2}$.

Rather than proceeding to derive a law such as (325), we illustrate an equivalent method for determining α^* that gives conceptual insight into the iteration, and that tends to be simpler for computation. In fact, the method we present here can be used to accelerate the computations of the preceding Section (11.1). Namely, the evolution laws take the form

$$E_{\alpha,(k+1)} \leq C_\alpha Z^{-f(\alpha)} e_{R,(k)}^{\delta f(\alpha)} E_{\alpha,(k)}, \tag{326}$$

where we observe that Z^{-1} and $e_{R,(k)}^{\delta}$ are raised to the same power $f(\alpha)$. This dimensional observation is a direct consequence of the Ansatz (269). The method we present in the following section allows us to determine whether these exponents are positive or negative, which allows us to determine whether choosing Z large can make such an evolution factor small at stage $(k) = 0$. Informally, the quantities whose evolution remains under control in the late stages of the iteration are also the quantities whose evolution we can control at the beginning of the iteration.

11.2.4 Asymptotics for the Parameters

In order to determine whether or not quantities such as the bounds

$$E_{\alpha,(k)} \equiv \left(\frac{e_v}{e_R}\right)_{(k)}^{\alpha/2} \Xi_{(k)}^{\alpha} e_{R,(k)}^{1/2-2\delta\alpha} \tag{327}$$

for the Hölder norms remain under control in the late stages of the iteration, it suffices to understand the asymptotic relationships

$$e_{v,(k)} \approx \Xi_{(k)}^{\beta_v} \tag{328}$$

$$e_{R,(k)} \approx \Xi_{(k)}^{\beta_R} \tag{329}$$

between the energy and frequency parameters that emerge during the late stages of the iteration. The following method can be used to determine these relationships, and reduces the control of quantities such as $E_{\alpha,(k)}$ to simple linear algebra calculations.

A Remark about Dimensional Analysis. The asymptotic relationships (328) and (329) cannot be deduced from dimensional analysis, since the dimensions of time and space for the Euler equations are independent of each other. Rather, these relationships depend on the efficiency of the convex integration procedure that we have constructed.

Heuristics for Parameter Evolution Let us recall the parameter evolution laws

$$e_{v,(k+1)} = e_{R,(k)} \tag{330}$$

$$e_{R,(k+1)} = Z^{-1} e_{R,(k)}^{1+\delta} \tag{331}$$

$$\Xi_{(k+1)} = CZ^2 \left(\frac{e_v}{e_R}\right)_{(k)}^{1/2} e_{R,(k)}^{-2\delta} \Xi_{(k)}. \tag{332}$$

In view of the formulas (327) and (332), it is more convenient to study the evolution of the energy ratio $\left(\frac{e_v}{e_R}\right)_{(k)}$, rather than the energy level $e_{v,(k)}$. For the energy ratio, we have the law

$$\left(\frac{e_v}{e_R}\right)_{(k+1)} = Z e_{R,(k)}^{-\delta}. \tag{333}$$

We can summarize these laws compactly by taking the logarithms of the equations, and writing

$$
\begin{bmatrix} \log e_R \\ \log(e_v/e_R) \\ \log \Xi \end{bmatrix}_{(k+1)} \approx \begin{bmatrix} 1+\delta & 0 & 0 \\ -\delta & 0 & 0 \\ -2\delta & 1/2 & 1 \end{bmatrix} \begin{bmatrix} \log e_R \\ \log(e_v/e_R) \\ \log \Xi \end{bmatrix}_{(k)} . \tag{334}
$$

This evolution law is not completely precise because it forgets about the constants in the exact parameter evolution laws; however, the **parameter evolution matrix**

$$
T_\delta = \begin{bmatrix} 1+\delta & 0 & 0 \\ -\delta & 0 & 0 \\ -2\delta & 1/2 & 1 \end{bmatrix} \tag{335}
$$

contains the most important information regarding the asymptotic relationships between the parameters. Informally, we have an approximation

$$
\begin{bmatrix} \log e_R \\ \log(e_v/e_R) \\ \log \Xi \end{bmatrix}_{(k)} \approx \begin{bmatrix} T_\delta^k \end{bmatrix} \begin{bmatrix} \log e_R \\ \log(e_v/e_R) \\ \log \Xi \end{bmatrix}_{(0)} . \tag{336}
$$

The reason is that, regardless of the initial choice of parameters, the right-hand side will point in the direction of the eigenvector of T_δ corresponding to its largest eigenvalue λ_+. In this case, T_δ is a lower triangular matrix whose eigenvalues 0, 1 and $\lambda_+ = 1 + \delta$ can be immediately read from the diagonal or by inspecting the form of the matrix. For example, the top row shows that $[1, 0, 0]$ is a row eigenvector for $\lambda_+ = (1 + \delta)$, the third column gives 1 as an eigenvalue, and the second and third columns are linearly dependent, giving 0 as another eigenvalue. Because $\lambda_+ > 1$, the heuristic (336) suggests that

$$
\begin{bmatrix} \log e_R \\ \log(e_v/e_R) \\ \log \Xi \end{bmatrix}_{(k)} \approx c_{(0)}(1+\delta)^k \psi_+ \tag{337}
$$

where ψ_+ is an eigenvector corresponding to the dominant eigenvalue

$$
T_\delta \psi_+ = (1+\delta)\psi_+
$$

which belongs to the **stable octant** $\mathcal{O}(-, +, +)$ defined by

$$
\psi_+ \in \mathcal{O}(-, +, +) = \left\{ \begin{bmatrix} a \\ b \\ c \end{bmatrix} \mid a < 0, b > 0, c > 0 \right\} \tag{338}
$$

of $\mathbb{R}^3$ in which the parameter logarithms

$$
\psi_{(k)} \equiv \begin{bmatrix} \log e_R \\ \log(e_v/e_R) \\ \log \Xi \end{bmatrix}_{(k)} \tag{339}
$$

eventually reside. The coefficient $c_{(0)}$ is obtained by expressing $\psi_{(0)}$ relative to a basis which diagonalizes T_δ, and can be assumed to be positive.

In this example, the minimal polynomial satisfied by the parameter evolution matrix is

$$(T_\delta - (1+\delta))T_\delta(T_\delta - 1) = 0. \tag{340}$$

Thus, one can take for ψ_+ any vector in $\operatorname{Im} T_\delta(T_\delta - 1)$ which also belongs to the stable octant $\mathcal{O}(-,+,+)$, such as

$$\psi_+ = -\delta^{-1}T_\delta(T_\delta - 1)\begin{bmatrix} 1 \\ 0 \\ 0 \end{bmatrix} \tag{341}$$

$$= -T_\delta \begin{bmatrix} 1 \\ -1 \\ -2 \end{bmatrix} \tag{342}$$

$$\psi_+ = \begin{bmatrix} -1 \\ 0 \\ \frac{5}{2} \end{bmatrix} + \delta \begin{bmatrix} -1 \\ 1 \\ 2 \end{bmatrix}. \tag{343}$$

Applying the heuristic (337) to the approximation

$$\log E_{\alpha,(k)} \approx (1/2 - 2\delta\alpha)\log e_{R,(k)} + \frac{\alpha}{2}\log(e_v/e_R)_{(k)} + \alpha\log \Xi_{(k)}$$
$$\log E_{\alpha,(k)} \approx [1/2 - 2\delta\alpha, \alpha/2, \alpha]\psi_{(k)}$$
$$\approx c_{(0)}(1+\delta)^k[1/2 - 2\delta\alpha, \alpha/2, \alpha]\psi_+$$
$$= c_{(0)}(1+\delta)^k[(1/2 - 2\delta\alpha)\cdot(-1-\delta) + (\alpha/2)\cdot(\delta) + (\alpha)\cdot(5/2 + 2\delta)]$$
$$= c_{(0)}(1+\delta)^k\left[-\frac{(1+\delta)}{2} + (\frac{5}{2} + \frac{9\delta}{2} + 2\delta^2)\alpha\right].$$

According to this approximation, the bounds on the Hölder norms decay at a double-exponential rate of

$$E_{\alpha,(k)} \approx e^{-c_\alpha(1+\delta)^k} \tag{344}$$

with $c_\alpha > 0$ for all

$$\alpha < \alpha^* = \frac{(1+\delta)}{5 + 9\delta + 4\delta^2}, \tag{345}$$

so that the resulting solution can have any Hölder regularity less than $1/5$.

Let us discuss how to make this heuristic argument more precise.

Rigorous Proof of Hölder Regularity From (330)–(332), the precise version of the parameter evolution rule (334) is given as follows.

$$
\begin{bmatrix} \log e_R \\ \log\left(\frac{e_v}{e_R}\right) \\ \log \Xi \end{bmatrix}_{(k+1)} = \begin{bmatrix} -\log Z \\ \log Z \\ \log(CZ^2) \end{bmatrix} + \begin{bmatrix} 1+\delta & 0 & 0 \\ -\delta & 0 & 0 \\ -2\delta & \frac{1}{2} & 1 \end{bmatrix} \begin{bmatrix} \log e_R \\ \log\left(\frac{e_v}{e_R}\right) \\ \log \Xi \end{bmatrix}_{(k)}
$$
(346)

We construct a basis that diagonalizes the matrix T_δ above by letting ψ_+ be the $(1 + \delta)$-eigenvector for T_δ given in Equation (343), and choosing $\psi_0 \in \mathrm{NS}(T_\delta)$ and $\psi_1 \in \mathrm{NS}(T_\delta - I)$ to be eigenvectors of T_δ for the eigenvalues 0 and 1 respectively.

The vector $\psi_{(k)}$ of parameter logarithms defined in Equation (339) then decomposes in this basis as

$$
\psi_{(k)} = c_{+,(k)}\psi_+ + c_{0,(k)}\psi_0 + c_{1,(k)}\psi_1, \tag{347}
$$

and the parameter evolution rule (346) becomes

$$
c_{+,(k+1)} = u_+ + (1 + \delta) \cdot c_{+,(k)} \tag{348}
$$
$$
c_{0,(k+1)} = u_0 + 0 \cdot c_{0,(k)} \tag{349}
$$
$$
c_{1,(k+1)} = u_1 + 1 \cdot c_{1,(k)}, \tag{350}
$$

where $[-\log Z, \log Z, \log(CZ^2)]^t = u_+\psi_+ + u_0\psi_0 + u_1\psi_1$.

The coefficients $c_{0,(k)} = u_0$ are constant in k, and the coefficients $c_{1,(k)}$ grow at most linearly according to the estimate

$$
|c_{1,(k)}| \le k|u_1| + |c_{1,(0)}|. \tag{351}
$$

Meanwhile, we claim that for large k, the sequence of coefficients $c_{+,(k)}$ are positive and exponentially growing. To prove this claim, observe that $[1, 0, 0]$ is a row-eigenvector of T_δ corresponding to the eigenvalue $(1+\delta)$; in other words, $[1, 0, 0][T_\delta - (1 + \delta)I] = 0$. As a consequence, $[1, 0, 0]$ remains invariant under the projection to the $(1 + \delta)$-eigenspace:

$$
[1, 0, 0] \left(\frac{T_\delta}{1+\delta}\right)\left(\frac{T_\delta - I}{\delta}\right) = [1, 0, 0]. \tag{352}
$$

Applying the above row vector to (347) and to the eigenvector decomposition of $[-\log Z, \log Z, \log(CZ^2)]^t$ allows us to extract the coefficients

$$
c_{+,(k)} = -\frac{\log e_{R,(k)}}{1 + \delta}, \quad u_+ = \frac{\log Z}{1 + \delta}. \tag{353}
$$

For sufficiently large k (say $k = k_0$), we have $e_{R,(k_0)} < 1$ and hence $c_{+,(k_0)} > 0$. For all $k \ge k_0$, we then have

$$
c_{+,(k)} \ge (1 + \delta)^{k-k_0} c_{+,(k_0)} > 0 \tag{354}
$$

by induction from Equation (348). Letting $E_{\alpha,(k)}$ be as defined in Equation (324), we have

$$\log E_{\alpha,(k)} = [1/2 - 2\delta\alpha, \alpha/2, \alpha]\psi_{(k)} \tag{355}$$

$$= c_{+,(k)}[1/2 - 2\delta\alpha, \alpha/2, \alpha]\psi_+ + O(|k|). \tag{356}$$

For all $\alpha < \alpha^*$ as in Equation (345), we have that $[1/2 - 2\delta\alpha, \alpha/2, \alpha]\psi_+ < 0$, which together with (354) implies that the above quantity is less than or equal to $-c_\alpha(1 + \delta)^k + O(|k|)$ for some $c_\alpha > 0$ and all $k \geq k_0$.

This calculation establishes for large k a super-exponential decay estimate on the bounds of the Hölder norms

$$E_{\alpha,(k)} = e^{-c_\alpha(1+\delta)^k + O(|k|)} \qquad \alpha < \alpha^*$$

for some constant $c_\alpha > 0$, and therefore concludes the proof of the Hölder regularity.

11.2.5   Regularity of the Pressure

Building on the calculations and methods of Sections (11.2.3) and (11.2.4), we can quickly determine what other quantities remain under control during the iteration.

For example, the pressure increment satisfies the estimates

$$\|\Gamma_{(k)}\|_{C^0} \leq Ce_{R,(k)} \tag{357}$$

$$\|\nabla P_{(k)}\|_{C^0} \leq CN_{(k)}\Xi_{(k)}e_{R,(k)} \tag{358}$$

$$\leq C\left(\frac{e_v}{e_R}\right)_{(k)}^{1/2} e_{R,(k)}^{-2\delta}\Xi_{(k)}e_{R,(k)} \tag{359}$$

and just as in the computations at the beginning of Section (11.2.3), writing the time derivative as

$$\partial_t = (\partial_t + v \cdot \nabla) - v \cdot \nabla$$

reveals that the time derivative of P obeys the same quality of estimates as the spatial derivatives, leading to the bounds

$$\|P_{(k)}\|_{C^0} \leq Ce_{R,(k)} \tag{360}$$

$$\|\nabla_{t,x}P_{(k)}\|_{C^0} \leq C\left(\frac{e_v}{e_R}\right)_{(k)}^{1/2} e_{R,(k)}^{-2\delta}\Xi_{(k)}e_{R,(k)} \tag{361}$$

which by interpolation give Hölder estimates of the form

$$\|P_{(k)}\|_{C_{t,x}^\alpha} \leq C\left(\frac{e_v}{e_R}\right)_{(k)}^{\alpha/2} \Xi_{(k)}^\alpha e_{R,(k)}^{1-2\delta\alpha}. \tag{362}$$

Following the methods of Section (11.2.4), define

$$E_{\alpha,(k)} \equiv \left(\frac{e_v}{e_R}\right)^{\alpha/2}_{(k)} \Xi^{\alpha}_{(k)} e^{1-2\delta\alpha}_{R,(k)} \tag{363}$$

so that $\|P_{(k)}\|_{C^{\alpha}_{t,x}} \leq C E_{\alpha,(k)}$ and

$$\log E_{\alpha,(k)} = [1 - 2\delta\alpha, \frac{\alpha}{2}, \alpha]\psi_{(k)}. \tag{364}$$

As in Section (11.2.4), we have that $E_{\alpha,(k)}$ decays super-exponentially when

$$[1 - 2\delta\alpha, \frac{\alpha}{2}, \alpha]\psi_+ = [1 - 2\delta\alpha, \frac{\alpha}{2}, \alpha]\left(\begin{bmatrix} -1 \\ 0 \\ \frac{5}{2} \end{bmatrix} + \delta \begin{bmatrix} -1 \\ 1 \\ 2 \end{bmatrix}\right) \tag{365}$$

is negative. In this case,

$$[1 - 2\delta\alpha, \frac{\alpha}{2}, \alpha]\psi_+ = -(1+\delta) + \alpha\left(\frac{5}{2} + (1+\delta)2\delta + \frac{\delta}{2} + 2\delta\right) \tag{366}$$

which is negative when

$$\alpha < 2\alpha^* = \frac{1+\delta}{5 + 9\delta + 4\delta^2}.$$

Thus, for the solution (v, p), $p \in C^{\alpha}_{t,x}$ when $\alpha < 2\alpha^*$.

11.2.6 Compact Support in Time

According to the choices made in Section (11.2.2), the time interval $T_{(k)}$ containing the support of the solutions grows by at most a multiple of the time scale $\theta_{(k)} = \Xi^{-1}_{(k)} e^{-1/2}_{v,(k)}$ after each iteration. Therefore, the resulting solution has compact support in time if the series

$$\sum_{(k)} \theta_{(k)} < \infty \tag{367}$$

converges. Note that the time scale $\theta^{-1}_{(k)}$ is essentially the Lipschitz norm of the approximate solution $v_{(k)}$. The fact that the natural time scale tends to 0 as we add in the finer scales $\Xi^{-1}_{(k)}$ is ultimately tied to the fact that the solution we construct is far from being a Lipschitz vector field.

To determine whether the series (367) is finite, it suffices to show that the logarithm of the time scale

$$\log \theta_{(k)} = -\frac{1}{2}\log(e_{v,(k)}) - \log \Xi_{(k)} \tag{368}$$

$$= -\frac{1}{2}\log e_{R,(k)} - \frac{1}{2}\log(e_v/e_R)_{(k)} - \log \Xi_{(k)} \tag{369}$$

decreases super-exponentially, and for this purpose it is enough to check that

$$[-1/2, -1/2, -1]\psi_+ = [-1/2, -1/2, -1]\left(\begin{bmatrix} -1 \\ 0 \\ \frac{5}{2} \end{bmatrix} + \delta \begin{bmatrix} -1 \\ 1 \\ 2 \end{bmatrix}\right) \quad (370)$$

$$= \frac{(1+\delta)}{2} - \frac{\delta}{2} - \left(\frac{5}{2} + 2\delta\right) < -2 \quad (371)$$

is negative, which is clearly the case.

Actually, one can even show that the time interval containing the solution can be made arbitrarily small by first observing that

$$\theta_{(0)} = \Xi_{(0)}^{-1} e_{v,(0)}^{-1/2} \quad (372)$$

$$= \Xi_{(0)}^{-1} [Z^{-\frac{1}{2(1+\delta)}} e_{R,(0)}^{-\frac{1}{2(1+\delta)}}] \quad (373)$$

can be made arbitrarily small by choosing Z large, and then using the evolution rules for the above formula to observe that one can arrange for

$$\theta_{(k+1)} \leq \frac{\theta_{(k)}}{2} \quad (374)$$

to hold for all $0 \leq k < \infty$ once Z is chosen large enough.

11.2.7 Nontriviality of the Solution

In order to show that the solutions produced by this process are nontrivial, we prove that the energy

$$\frac{1}{2} \int |v_{(k)}|^2 dx$$

of $v_{(k)} = \sum_{(j)=0}^{k} V_{(j)}$ eventually increases in k on the nonempty interval $I_{(0)}$.

Suppose that $t^* \in I_{(0)}$. To verify that $\frac{1}{2} \int |v_{(k)}|^2 (t^*, x) dx$ eventually increases with (k), we calculate

$$\int \frac{|v_{(k+1)}|^2}{2}(t^*, x)dx - \int \frac{|v_{(k)}|^2}{2}(t^*, x)dx$$

$$= \int \left[\frac{|v_{(k)} + V_{(k)}|^2}{2} - \frac{|v_{(k)}|^2}{2}\right]dx \quad (375)$$

$$= \int \frac{|V_{(k)}|^2}{2}dx + \int v_{(k)} \cdot V_{(k)} dx. \quad (376)$$

Our choice of $e_{(k)}$ assures us that we have a lower bound of

$$\int \frac{|V_{(k)}|^2}{2}(t^*, x)dx \geq A e_{R,(k)} \quad (377)$$

for some constant A, since Lemma (10.1) implies that the difference

$$\int \frac{|V_{(k)}|^2}{2} dx - \int e_{(k)}(t)dx$$

is bounded by

$$|\int \frac{|V_{(k)}|^2}{2}(t,x)dx - \int e_{(k)}(t)dx| \le C\frac{e_{R,(k)}}{N_{(k)}} \tag{378}$$

and $N_{(k)}$ is an increasing sequence whose first term

$$N_{(0)} = Z^2 \left(\frac{e_v}{e_R}\right)_{(0)}^{1/2} e_{R,(0)}^{-2\delta} \tag{379}$$

$$= Z^{2+\frac{\delta}{2(1+\delta)}} e_{R,(0)}^{-\delta(2+\frac{1}{2(1+\delta)})} \tag{380}$$

can be made arbitrarily large by choosing Z large.

Thus, in order to check that the energy increases for each stage (k), it suffices to bound the inner product

$$<v_{(k)}, V_{(k)}>_{L^2(\mathbb{T}^3)} = \int v_{(k)} \cdot V_{(k)} dx \tag{381}$$

which measures the correlation between the correction $V_{(k)}$ and the velocity field $v_{(k)}$.

We expect the inner product (381) to be fairly small because $V_{(k)}$ oscillates at a much higher frequency than $v_{(k)}$. Thus, the first step to estimating the inner product (381) is to write

$$V_{(k)} = \nabla \times W_{(k)}$$

and integrate by parts to see that

$$<v_{(k)}, V_{(k)}> = \int v_{(k)} \cdot \nabla \times W_{(k)} dx \tag{382}$$

$$= \int \nabla \times v_{(k)} \cdot W_{(k)} dx \tag{383}$$

which, according to Lemma (10.1) and $\|\nabla v_{(k)}\|_{C^0} \le \Xi_{(k)} e_{v,(k)}^{1/2}$ from the definition of frequency and energy levels, implies a bound

$$|<v_{(k)}, V_{(k)}>| \le C[\Xi_{(k)} e_{v,(k)}^{1/2}] \cdot [\frac{e_{R,(k)}^{1/2}}{\Xi_{(k)} N_{(k)}}] \tag{384}$$

$$\le C\frac{e_{v,(k)}^{1/2} e_{R,(k)}^{1/2}}{N_{(k)}} \tag{385}$$

$$= C\left(\frac{e_v}{e_R}\right)_{(k)}^{1/2} N_{(k)}^{-1} e_{R,(k)} \tag{386}$$

$$= C(Z^{-2} e_{R,(k)}^{2\delta}) e_{R,(k)}. \tag{387}$$

Applying this estimate to the formula (376) for the energy increment, we can even ensure using

$$\int \frac{|v_{(k+1)}|^2}{2}(t^*,x)dx - \int \frac{|v_{(k)}|^2}{2}(t^*,x)dx \geq A e_{R,(k)} - C(Z^{-2}e_{R,(k)}^{2\delta})e_{R,(k)} \tag{388}$$

that the energy increment is positive at the initial stage $(k) = 0$ by choosing Z large. Since $e_{R,(k)}^{2\delta}$ decreases, the energy increment will be positive at all future stages as well.

Thus, the solution obtained at the end of the process is nontrivial because the energy increases at t^* after each iteration, and because $e_{R,(0)}$ was arbitrary, we can make the energy of the solutions arbitrarily large.

This observation completes the proof of Theorem (10.1).

$\square$

12 Gluing Solutions

Here we note that the method of the previous section can be extended to prove Theorem (0.2) on the gluing of solutions, which states the following.

Theorem 12.1. *For every smooth solution (v, p) to incompressible Euler on $(-2, 2) \times \mathbb{T}^3$, there exists a Hölder continuous solution $(\bar{v}, \bar{p})$ which coincides with (v, p) on $(-1, 1) \times \mathbb{T}^3$ but is equal to a constant outside of $(-3/2, 3/2) \times \mathbb{T}^3$.*

Proof. By taking a Galilean transformation if necessary, we can assume without loss of generality that the solution has zero total momentum $\int v^l(t, x)dx = 0$. Let $\bar{\eta}(t)$ be a smooth cutoff function equal to 1 in a neighborhood of $[-1, 1]$, and equal to 0, outside $(-6/5, 6/5)$. Then the cut off velocity and pressure $v_{(0)}(t, x) = \bar{\eta}(t)v(t, x)$, $p_{(0)}(t, x) = \bar{\eta}(t)p$ form a smooth solution to the Euler-Reynolds equations with compact support when coupled to a smooth stress tensor $R_{(0)}^{jl}$ that solves

$$\partial_j R_{(0)}^{jl} = \bar{\eta}'(t)v^l + \partial_j[(\bar{\eta}^2(t) - \bar{\eta}(t))v^j v^l]. \tag{389}$$

The existence of a smooth R^{jl} solving (389) follows from the fact that the right-hand side has integral 0.

The proof of Theorem (12.1) then proceeds by iterating Lemma (10.1) just as in the proof of Theorem (10.1). Here the only difference is that the sets $I_{(k)}$ containing the support of each $R_{(k)}$ are unions of two intervals $I = I^1 \cup I^2$, with $I_{(k)}^1 \subseteq (-\infty, 1)$ and $I_{(k)}^2 \subseteq (1, \infty)$.

We still choose energy functions of the form

$$e_{(k)}^{1/2}(t) = (20K e_{R,(k)})^{1/2}\eta_{\tau_{(k)}} * \chi_{(k)}(t) \tag{390}$$

with

$$\chi_{(k)}(t) = \chi_{I_{(k)} \pm (\theta_{(k)} + \tau_{(k)})}(t)$$

being the characteristic function of a slight expansion of $I_{(k)}$. According to the computations in Section (11.2.6), we can ensure that the corrections, which live on small neighborhoods of the supports of $R_{(k)}$, never touch the intervals $(-\infty, -3/2] \cup [-1, 1] \cup [-3/2, \infty)$ by taking the initial frequency level $\Xi_{(0)}$ for $(v_{(0)}, p_{(0)}, R_{(0)})$ to be sufficiently large. Applying another Galilean transformation to return to the original frame of reference, we have the theorem. $\qquad\square$

13   On Onsager's Conjecture

As mentioned at the end of Section (10), a stronger form of Lemma (10.1) would imply Onsager's conjecture. Here we investigate what could be proven assuming Conjecture (10.1).

Theorem 13.1. *Assume Conjecture (10.1). Then for every $\delta > 0$, there exist nontrivial weak solutions (v, p) to the Euler equations on $\mathbb{R} \times \mathbb{T}^3$ such that for*

$$\alpha^* = \frac{1}{3 + 2\delta}$$

we have

$$v \in C_{t,x}^{\alpha}(\mathbb{R} \times \mathbb{R}^3) \qquad \text{for all } \alpha < \alpha^* \tag{391}$$

$$p \in C_{t,x}^{2\alpha}(\mathbb{R} \times \mathbb{R}^3) \tag{392}$$

whose support is contained in a bounded time interval.
    Furthermore, one can arrange that the energy

$$E(t) = \frac{1}{2} \int |v|^2(t, x) dx \tag{393}$$

is C^1 as a function of t. In particular, the energy will increase or decrease in certain time intervals.

Remark. In fact, the proof shows that, at least for the solutions constructed by this method, the energy $E(t)$ will necessarily belong to $C^{0,\alpha}$ for all $\alpha < 1$.

Proof. We iterate the conjectured Lemma (11) so that the energy levels

$$\frac{e_{v,(k)}^{1/2} e_{R,(k)}^{1/2}}{N_{(k)}} = e_{R,(k+1)} = \frac{e_R^{1+\delta}}{Z} \tag{394}$$

decay super-exponentially. This Ansatz leads to a choice of

$$N_{(k)} = Z \left(\frac{e_v}{e_R} \right)_{(k)}^{1/2} e_{R,(k)}^{-\delta} \tag{395}$$

$$\Xi_{(k+1)} = CZ \left(\frac{e_v}{e_R} \right)_{(k)}^{1/2} e_{R,(k)}^{-\delta} \Xi_{(k)} \tag{396}$$

$$\left(\frac{e_v}{e_R} \right)_{(k+1)} = Z e_{R,(k)}^{-\delta} \tag{397}$$

resulting in a parameter evolution matrix

$$T_\delta = \begin{bmatrix} 1+\delta & 0 & 0 \\ -\delta & 0 & 0 \\ -\delta & 1/2 & 1 \end{bmatrix} \tag{398}$$

for the parameter logarithms

$$\psi_{(k)} = \begin{bmatrix} \log e_R \\ \log(e_v/e_R) \\ \log \Xi \end{bmatrix}_{(k)} . \tag{399}$$

Since T_δ satisfies

$$(T_\delta - (1+\delta))T_\delta(T_\delta - 1) = 0 \tag{400}$$

the stable direction for iterating T_δ corresponds to the portion of the $1+\delta$ eigenspace of T_δ in the stable octant

$$\mathcal{O}(-,+,+) = \left\{ \begin{bmatrix} a \\ b \\ c \end{bmatrix} \mid a < 0, b > 0, c > 0 \right\} \tag{401}$$

in which the parameter logarithms $\psi_{(k)}$ eventually reside.

To find a vector in $\mathcal{O}$ belonging to the $1+\delta$ eigenspace, we compute

$$\psi_+ = -\delta^{-1} T_\delta(T_\delta - 1) \begin{bmatrix} 1 \\ 0 \\ 0 \end{bmatrix} \tag{402}$$

$$= -T_\delta \begin{bmatrix} 1 \\ -1 \\ -1 \end{bmatrix} \tag{403}$$

$$\psi_+ = \begin{bmatrix} -1 \\ 0 \\ \frac{3}{2} \end{bmatrix} + \delta \begin{bmatrix} -1 \\ 1 \\ 1 \end{bmatrix} . \tag{404}$$

In order to determine which Hölder norms stay under control during the iteration, we again observe that the bound for the spatial derivative of the corrections V and P also controls their full space-time derivative since

$$\partial_t = (\partial_t + v \cdot \nabla) - v \cdot \nabla$$

and the material derivatives obey even better estimates. Therefore we have an estimate

$$\|V_{(k)}\|_{C^\alpha_{t,x}} \le C(N_{(k)}\Xi_{(k)})^\alpha e_{R,(k)}^{1/2} \tag{405}$$

$$\le C'\Xi_{(k)}^\alpha \left(\frac{e_v}{e_R}\right)_{(k)}^{\alpha/2} e_{R,(k)}^{1/2-\delta\alpha} \tag{406}$$

$$\|P_{(k)}\|_{C^\alpha_{t,x}} \le C(N_{(k)}\Xi_{(k)})^\alpha e_{R,(k)} \tag{407}$$

$$\le C'\Xi_{(k)}^\alpha \left(\frac{e_v}{e_R}\right)_{(k)}^{\alpha/2} e_{R,(k)}^{1-\delta\alpha}. \tag{408}$$

As in Section (11), we take the logarithm of (406) to see that the solution has Hölder regularity $v \in C^\alpha_{t,x}$ if

$$[(1/2 - \delta\alpha), \alpha/2, \alpha]\left(\begin{bmatrix} -1 \\ 0 \\ \frac{3}{2} \end{bmatrix} + \delta \begin{bmatrix} -1 \\ 1 \\ 1 \end{bmatrix}\right)$$
$$= -\frac{(1+\delta)}{2} + \delta(1+\delta)\alpha + \frac{\delta\alpha}{2} + \left(\frac{3\alpha}{2} + \delta\alpha\right) \tag{409}$$

$$= -\frac{(1+\delta)}{2} + \frac{\alpha}{2}(3 + 2\delta)(1+\delta) \tag{410}$$

is negative. Therefore, the norm whose bounds grow exponentially corresponds to the regularity $\alpha^* = \frac{1}{3+2\delta}$.

Similarly, we see that $p \in C^\alpha_{t,x}$ for $\alpha < 2\alpha^*$ by taking a logarithm and computing

$$[(1 - \delta\alpha), \alpha/2, \alpha]\left(\begin{bmatrix} -1 \\ 0 \\ \frac{3}{2} \end{bmatrix} + \delta \begin{bmatrix} -1 \\ 1 \\ 1 \end{bmatrix}\right)$$
$$= -(1+\delta) + \frac{\alpha}{2}(3 + 2\delta)(1+\delta).$$

13.1 Higher Regularity for the Energy

Here we observe that the energy

$$E(t) = \int |v|^2(t,x)dx \tag{411}$$

92

of the solution v that we ultimately construct actually enjoys better regularity than the solution v itself has in the time variable. Namely, the construction necessarily results in a v for which $E(t) \in C_t^\alpha$ for all $\alpha < 1$, and by choosing $e_{(k)}$ carefully at each stage, we can even arrange that $E(t) \in C^1(\mathbb{R})$.

Let us begin by writing the energy as a sum of energy increments

$$E(t) = \sum_{k=0}^{\infty} E_{(k)} \tag{412}$$

$$E_{(k)}(t) = \int \left(\frac{|v_{(k+1)}|^2}{2}(t,x) - \frac{|v_{(k)}|^2}{2}(t,x) \right) dx \tag{413}$$

$$= \int \frac{|V_{(k)}|^2}{2} dx + \int v_{(k)} \cdot V_{(k)} dx. \tag{414}$$

For the ideal case with $\mathbf{b}^{-1}$, the Main Lemma, Lemma (10.1), implies that

$$\left\| \int_{\mathbb{T}^3} |V_{(k)}|^2 dx - \int_{\mathbb{T}^3} e_{(k)}(t) dx \right\|_{C^0} \leq C \frac{e_{R,(k)}}{N_{(k)}} \tag{415}$$

$$\left\| \frac{d}{dt} \left(\int_{\mathbb{T}^3} |V_{(k)}|^2 dx - \int_{\mathbb{T}^3} e_{(k)}(t) dx \right) \right\|_{C^0} \leq C \frac{\Xi_{(k)} e_{v,(k)}^{1/2} e_{R,(k)}}{N_{(k)}}. \tag{416}$$

In fact, for the ideal case, the quantity

$$\Xi_{(k)} e_{v,(k)}^{1/2} e_{R,(k)}$$

is exactly the critical quantity which grows exponentially, in the sense that

$$\Xi_{(k+1)} e_{v,(k+1)}^{1/2} e_{R,(k+1)} = C N_{(k)} \Xi_{(k)} e_{v,(k)}^{1/2} e_{R,(k)} \frac{e_{v,(k)}^{1/2} e_{R,(k)}^{1/2}}{N_{(k)}} \tag{417}$$

$$= C \Xi_{(k)} e_{v,(k)}^{1/2} e_{R,(k)} \tag{418}$$

Since $N_{(k)}$ grows super-exponentially, we see from (416) that the error for prescribing the energy of the correction can actually be summed in $C^1(\mathbb{R})$ for any admissible choices of energy functions. Namely, from (414), we have

$$\frac{d}{dt} E_{(k)}(t) = \frac{d}{dt} \int e(t) dx + \frac{d}{dt} \int v_{(k)} \cdot V_{(k)} dx + o(1) \tag{419}$$

where $o(1)$ goes to 0 in C_t^0 uniformly in (k).

We can also sum the correlation term

$$\left\| \frac{d}{dt} \int v_{(k)} \cdot V_{(k)} dx \right\|_{C_t^0} \leq C \frac{\Xi_{(k)} e_{v,(k)} e_{R,(k)}^{1/2}}{N_{(k)}} \tag{420}$$

$$\leq C \Xi_{(k)} e_{v,(k)}^{1/2} e_{R,(k)}^{1+\delta} \tag{421}$$

$$= o(1) \tag{422}$$

which implies that

$$\|\frac{d}{dt}E_{(k)}(t) - \frac{d}{dt}\int e_{(k)}(t)dx\|_{C_t^0} = o(1) \tag{423}$$

where $e_{(k)}(t)$ is the energy function chosen at stage (k).

To establish the estimate (420), we write

$$\frac{d}{dt}\int v_{(k)} \cdot V_{(k)}dx = \frac{d}{dt}\int <v_{(k)}, \nabla \times W_{(k)}>dx \tag{424}$$

$$= \frac{d}{dt}\int <\nabla \times v_{(k)}, W_{(k)}>dx \tag{425}$$

$$= \int (\partial_t + v_{(k)}^j\partial_j)<\nabla \times v_{(k)}, W_{(k)}>dx \tag{426}$$

$$= \int <(\partial_t + v_{(k)}^j\partial_j)\nabla \times v_{(k)}, W_{(k)}>dx$$
$$+ \int <\nabla \times v_{(k)}, (\partial_t + v_{(k)}^j\partial_j)W_{(k)}>dx. \tag{427}$$

The second of these terms can be estimated using the bounds for $W_{(k)}$ in Lemma (10.1).

$$\left\|\int_{\mathbb{T}^3} <\nabla \times v_{(k)}, (\partial_t + v_{(k)}^j\partial_j)W_{(k)}>dx\right\|_{C_t^0} \tag{428}$$
$$\leq C\|\nabla v_{(k)}\|_{C^0}\|(\partial_t + v_{(k)}^j\partial_j)W_{(k)}\|_{C^0}$$

$$\leq C\left(\Xi_{(k)}e_{v,(k)}^{1/2}\right)\left(\frac{e_{v,(k)}^{1/2}e_{R,(k)}^{1/2}}{N_{(k)}}\right) \tag{429}$$

$$\leq C(\Xi_{(k)}e_{v,(k)}^{1/2})e_{R,(k)}^{1+\delta} = o(1) \tag{430}$$

The first term can be estimated by commuting the curl $\nabla\times$ and the material derivative $\partial_t + v \cdot \nabla$.

$$\int <(\partial_t + v_{(k)}^j\partial_j)\nabla \times v_{(k)}, W_{(k)}>dx$$
$$= \int <\nabla \times (\partial_t + v_{(k)} \cdot \nabla)v_{(k)}, W_{(k)}>dx + \int <u_{(k)}, W_{(k)}>dx$$
$$= E_{1,(k)} + E_{2,(k)} \tag{431}$$
$$u_{(k)} = (\partial_t + v_{(k)} \cdot \nabla)\nabla \times v_{(k)} - \nabla \times (\partial_t + v_{(k)} \cdot \nabla)v_{(k)} \tag{432}$$

Using the definition $(\nabla \times w)^l = \epsilon^l{}_{ab}\partial^a w^b$, we can compute the commutator $u_{(k)}$

$$(\nabla \times (\partial_t + v_{(k)} \cdot \nabla)v_{(k)})^l = \epsilon^l{}_{ab}\partial^a[(\partial_t + v^j_{(k)}\partial_j)v^b_{(k)}] \tag{433}$$

$$= (\partial_t + v^j_{(k)}\partial_j)\epsilon^l{}_{ab}\partial^a v^b_{(k)} + \epsilon^l{}_{ab}\partial^a v^j_{(k)}\partial_j v^b_{(k)} \tag{434}$$

$$u^l_{(k)} = -\epsilon^l{}_{ab}\partial^a v^j_{(k)}\partial_j v^b_{(k)} \tag{435}$$

allowing us to bound the term $E_{2,(k)}$ by comparing it to the exponential growth of $\Xi_{(k)}e^{1/2}_{v,(k)}e_{R,(k)}$

$$\left\| \int <u_{(k)}, W_{(k)}>dx \right\|_{C^0_t} \le \|u_{(k)}\|_{C^0_t L^{3/2}_x}\|W_{(k)}\|_{C^0_t L^3_x}$$

$$\le C\|\nabla v_{(k)}\|^2_{C^0_t L^3_x}\|W_{(k)}\|_{C^0_t L^3_x}$$

$$\le C\left(\Xi_{(k)}e^{1/2}_{v,(k)}\right)^2 \left(\Xi^{-1}_{(k)}N^{-1}_{(k)}e^{1/2}_{R,(k)}\right)$$

$$\le C\Xi_{(k)}e^{1/2}_{v,(k)}\frac{e^{1/2}_{v,(k)}e^{1/2}_{R,(k)}}{N_{(k)}}$$

$$\le C\Xi_{(k)}e^{1/2}_{v,(k)}e^{1+\delta}_{R,(k)}.$$

To bound the other term $E_{1,(k)}$, we can use the Euler-Reynolds equation

$$(\partial_t + v_{(k)} \cdot \nabla)v^l_{(k)} = \partial^l p_{(k)} + \partial_j R^{jl}_{(k)} \tag{436}$$

to estimate

$$\left\| \int <\nabla \times (\partial_t + v_{(k)} \cdot \nabla)v_{(k)}, W_{(k)}>dx \right\|_{C^0_t} \tag{437}$$

$$\le (\|\nabla^2 p\|_{C^0} + \|\nabla^2 R\|_{C^0})\|W_{(k)}\|_{C^0}$$

$$\le C\Xi^2_{(k)}e_{v,(k)}\frac{e^{1/2}_{R,(k)}}{\Xi_{(k)}N_{(k)}} \tag{438}$$

$$\le C\Xi_{(k)}e^{1/2}_{v,(k)}e^{1+\delta}_{R,(k)} = o(1). \tag{439}$$

We have therefore established that

$$\left\| \frac{d}{dt}E_{(k)}(t) - \frac{d}{dt}\int e_{(k)}(t)dx \right\|_{C^0_t} = o(1). \tag{440}$$

If we desire the energy of the resulting solution to be C^1 in time, we must be more careful than using the bounds

$$\left\| \frac{d}{dt}e_{(k)}(t) \right\|_{C^0} = 2\|e^{1/2}_{(k)}(t)\frac{d}{dt}e^{1/2}_{(k)}(t)\|_{C^0} \tag{441}$$

$$\le C\Xi_{(k)}e^{1/2}_{v,(k)}e_{R,(k)} \tag{442}$$

$$\le C^k \tag{443}$$

required by Lemma (10.1), since these bounds grow exponentially.

However, we can easily choose $e_{(k)}$ so that the bounds (443) are not sharp.

For example, averaging over longer time intervals

$$\tau_{(k)} = B_0^k \Xi_{(k)}^{-1} e_{v,(k)}^{-1/2} \tag{444}$$

for the time mollifier in (307), (310) will lead to bounds which decay exponentially for

$$\|\frac{d}{dt} e_{(k)}(t)\|_{C^0} \leq C A^{-k} \tag{445}$$

if B_0 is chosen sufficiently large. $\qquad\square$

Part V

Construction of Regular Weak Solutions: Preliminaries

14 Preparatory Lemmas

To prepare for the proof, we present the following, well-known method concerning the general rate of convergence of mollifiers.

Lemma 14.1. *Suppose that* $(1 + |y|^N)\eta(y) \in L^1(\mathbb{R}^3)$ *and that for any multi-index* α *with* $1 \leq |\alpha| \leq N - 1$ *the moment vanishing conditions*

$$\int \eta(y)dy = 1 \tag{446}$$

$$\int y^\alpha \eta(y)dy = 0 \qquad \forall \alpha, 1 \leq |\alpha| \leq N - 1 \tag{447}$$

hold.

Let $\epsilon > 0$ *and define* $\eta_\epsilon(y) = \epsilon^{-3}\eta(\frac{y}{\epsilon})$. *Then there exists a constant* C *such that*

$$||v - \eta_\epsilon * v||_{C^0(\mathbb{T}^3)} \leq C\epsilon^N ||\eta||_{L^N(\mathbb{R}^3)}||\nabla^N v||_{C^0(\mathbb{T}^3)} \tag{448}$$

for all smooth functions v *on* $\mathbb{T}^3$.

Proof. To reduce the number of minus signs appearing in the proof, let us assume η_ϵ is even, and let us restrict to the case $N = 3$ to convey the idea. To verify (448), we can then Taylor expand by repeated integration

97

by parts in the formula

$$\eta_\epsilon * v(x) - v(x) = \int_{\mathbb{R}^3} (v(x - y) - v(x))\eta_\epsilon(y)dy$$

$$= \int (v(x + y) - v(x))\eta_\epsilon(y)dy$$

$$= \int (v(x + \epsilon y) - v(x))\eta(y)dy$$

$$= \int \left[\int_0^1 \frac{d}{ds}v(x + s\epsilon y)ds\right]\eta(y)dy$$

$$= \epsilon \int_0^1 \left[\int \partial_i v(x + s\epsilon y)y^i \eta(y)dy\right]ds$$

$$= \epsilon \int \partial_i v(x)y^i \eta(y)dy$$

$$+ \int \left[\int_0^1 \frac{d^2}{ds^2}v(x + s\epsilon y)(1 - s)ds\right]\eta(y)dy$$

$$= \epsilon^2 \int_0^1 \left[\int \partial_{i_2}\partial_{i_1} v(x + s\epsilon y)y^{i_1}y^{i_2}\eta(y)dy\right](1 - s)ds$$

$$= \frac{\epsilon^2}{2!} \int \partial_{i_2}\partial_{i_1} v(x)y^{i_1}y^{i_2}\eta(y)dy$$

$$+ \int \left[\int_0^1 \frac{d^3}{ds^3}v(x + s\epsilon y)\frac{(1 - s)^2}{2!}ds\right]\eta(y)dy$$

$$= \frac{\epsilon^3}{2!} \int_0^1 \left[\int \partial_{i_1}\partial_{i_2}\partial_{i_3} v(x + s\epsilon y)y^{i_1}y^{i_2}y^{i_3}\eta(y)dy\right](1 - s)^2 ds.$$

$\square$

Remark. The case $N = 2$ of the above lemma is used in [CDLS12b]. In their argument, they take advantage of how the case $N = 2$ permits η_ϵ to be non-negative. Actually, the case $N = 2$ also suffices for the proof of the main theorem in the present book, but we include the more general statement (448) to convey the flexibility available in this aspect of the argument.

Another way to work out the above details, which will be repeatedly used in the remainder of the proof, is to write

$$\eta_\epsilon * v(x) - v(x) = \int [v(x - y) - v(x)]\eta_\epsilon(y)dy \tag{449}$$

$$= -\int_0^1 \left[\int \partial_i v(x - sy)y^i \eta_\epsilon(y)dy\right]ds \tag{450}$$

$$= \epsilon \int_0^1 \int \partial_i v(x - sy)\tilde{\eta}_\epsilon^i(y)dyds \tag{451}$$

where

$$\tilde{\eta}_\epsilon^i(y) = -\frac{1}{\epsilon^3}\frac{y^i}{\epsilon}\eta(\frac{y}{\epsilon}) \tag{452}$$

are other functions whose integrals are not normalized to 1, but which satisfy the same type of estimates as η_ϵ.

Namely, every function η_ϵ and $\tilde{\eta}_\epsilon$ which appears in the argument will satisfy bounds of the form

$$\|\nabla^k \eta_\epsilon\|_{L^1(\mathbb{R}^3)} \leq C_\eta \epsilon^{-k} \tag{453}$$

and altogether through the course of the argument, only finitely many different η and $\tilde{\eta}$ will be used.

15   The Coarse Scale Velocity

To begin the proof of Lemma (10.1), we choose a mollification v_ϵ^j for the velocity field. For our purposes, it is convenient to take a double mollification

$$\eta_{\epsilon+\epsilon} - \eta_{\epsilon_v} * \eta_{\epsilon_v} \tag{454}$$

$$v_\epsilon^j - \eta_{\epsilon+\epsilon} * v^j \tag{455}$$

for reasons which will be explained in the footnote of Section (16) after we derive a transport equation for v_ϵ.

We would like to take ϵ_v to be as large as possible so that higher derivatives of v_ϵ are less costly. More specifically, the basic building blocks of the construction are of the form

$$e^{i\lambda\xi_I} v_I^l$$

and we must ensure that each v_I has frequency smaller than λ, so ϵ_v^{-1} must be smaller than λ in order of magnitude.

To achieve a frequency of $\Xi' = C\Xi N$ as in the statement of the Main Lemma, Lemma (10.1), we will take λ to have the form

$$\lambda = B_\lambda \Xi N \tag{456}$$

where $B_\lambda \geq 1$ is a very large constant which will be chosen at the very end of the proof.

The parameter ϵ_v will then have the form

$$\epsilon_v = a_v \Xi^{-1} N^{-\gamma} \tag{457}$$

$$0 < \gamma < 1 \tag{458}$$

where $a_v \leq 1$ is a small constant that will also be chosen in the proof to depend only on L. At this point we can discuss the exponent γ.

99

Analogously to Section (8.1) in the continuous case, we choose ϵ_v so that the main term in the error

$$(v^j - v_\epsilon^j)V^l + V^j(v^l - v_\epsilon^l) = \sum_I 2[(v - v_\epsilon)(e^{i\lambda\xi_I}v_I)]^{jl} + O(\lambda^{-1}) \quad (459)$$

$$= Q_v^{jl} + O(\lambda^{-1}) \quad (460)$$

will be small. Namely, we know a priori, as in (8.1), that each v_I will be no larger than

$$\|v_I\|_{C^0} \le Ae_R^{1/2} \quad (461)$$

for some absolute constant provided that we require

$$\tau \le b_0\Xi^{-1}e_v^{-1/2} \quad (462)$$

for a certain constant b_0 discussed in Section (8.1). Furthermore, there is a uniform bound on the number of v_I which are nonzero at any given (t,x), so we also have an a priori bound

$$\|\sum_I |v_I|\|_{C^0} \le Ae_R^{1/2}. \quad (463)$$

Therefore, we have an a priori bound on

$$\|Q_v\|_{C^0} \le Ae_R^{1/2}\|v - v_\epsilon\|_{C^0}. \quad (464)$$

In the ideal case for which there are no losses of $\mathbf{b}^{-1}$ in the Main Lemma (10.1), the target size for the new stress would be $\|R_1\|_{C^0} \lhd \frac{e_v^{1/2}e_R^{1/2}}{N}$. We will therefore choose ϵ_v to achieve the target

$$\|Q_v\|_{C^0} \lhd \frac{1}{50}\frac{e_v^{1/2}e_R^{1/2}}{N}. \quad (465)$$

To approach the target (465), let η_{ϵ_v} be any smooth mollifier constructed from an even function of compact support which satisfies the vanishing moment conditions of Lemma (446) for $N = L$. We then have an estimate

$$\|\eta_{\epsilon+\epsilon} * v - v\|_{C^0} = \|(\eta_{\epsilon_v} * v - v) + \eta_{\epsilon_v} * (\eta_{\epsilon_v} * v - v)\|_{C^0} \quad (466)$$

$$\le C\epsilon_v^L\|\nabla^L v\|_{C^0} \quad (467)$$

$$\le C\epsilon_v^L(\Xi^L e_v^{1/2}). \quad (468)$$

In anticipation of proving (465), we set $\gamma = 1/L$, so that

$$\epsilon_v = a_v\Xi^{-1}N^{-1/L} \quad (469)$$

and (468) becomes

$$\|\eta_{\epsilon+\epsilon} * v - v\|_{C^0} \le Ca_v^L\frac{e_v^{1/2}}{N}. \quad (470)$$

Applying the above inequality to (464) gives a bound of

$$\|Q_v\|_{C^0} \le A a_v^L e_R^{1/2} \frac{e_v^{1/2}}{N} \tag{471}$$

for the leading term in the stress created by mollifying v.

By comparing (471) with (465), we can choose the constant a_v so that the goal (465) is satisfied.

With the above choice, ∇v_ϵ has amplitude $\Xi e_v^{1/2}$ and frequency Ξ to high order

$$\|\partial^\alpha v_\epsilon\|_{C^0} \le \Xi^{|\alpha|} e_v^{1/2} \qquad \alpha = 1, \ldots, L \tag{472}$$

$$\|\partial^\alpha v_\epsilon\|_{C^0} \le C_\alpha N^{(|\alpha|-L)/L} \Xi^{|\alpha|} e_v^{1/2} \qquad |\alpha| > L \tag{473}$$

which we choose to abbreviate by

$$\|\partial^\alpha v_\epsilon\|_{C^0} \le C_\alpha N^{(|\alpha|-L)_+/L} \Xi^{|\alpha|} e_v^{1/2} \qquad \text{for all } |\alpha| \ge 1. \tag{474}$$

We can summarize (474) informally by saying that each derivative of v_ϵ up to order L costs a factor of Ξ, and then derivatives beyond order L cost $N^{1/L}\Xi$. The factor $N^{1/L}$ is like an "interest payment" which must be paid on top of the usual cost of Ξ for the "borrowed" derivatives beyond order L.

Note that we do not assume any control over $\|v_\epsilon\|_{C^0}$; in fact the low frequency part of v, though bounded, can be very large when the lemma is used in the proof of Theorem (10.1). There, the largest part of the C^0 norm comes from the very first step of the iteration.

At this point, we can already check to make sure that we have not fallen short of the goal

$$\|\nabla Q_v\|_{C^0} \trianglelefteq \Xi' e_{R'} \tag{475}$$

$$= C(N\Xi) \frac{e_v^{1/2} e_R^{1/2}}{N} \tag{476}$$

$$= C\Xi e_v^{1/2} e_R^{1/2} \tag{477}$$

by estimating

$$\|\nabla(v - v_\epsilon)\|_{C^0} \le \|\nabla v\|_{C^0} + \|\nabla v_\epsilon\|_{C^0} \tag{478}$$

$$\le 2\Xi e_v^{1/2}. \tag{479}$$

Since we expect that $|V| \lesssim e_R^{1/2}$, the crude estimate (479) will suffice to prove (475).

The Choice of ϵ_v and the Parametrix Expansion. It is crucial to observe that since $L \ge 2$, the choice of

$$\epsilon_v^{-1} \lesssim \Xi N^{1/L}$$

is significantly smaller than $\lambda \approx \Xi N$. We use this comparison in the parametrix expansion for the divergence equation, which proceeds by solving

$$\partial_j Q^{jl} = e^{i\lambda \xi_I} u_1^l \tag{480}$$

$$Q^{jl} = \tilde{Q}_1^{jl} + Q_{(1)}^{jl} \tag{481}$$

$$\tilde{Q}_1^{jl} = e^{i\lambda \xi_I} \frac{q^{jl}(\nabla \xi)[u^l]}{\lambda} \tag{482}$$

$$\partial_j Q_{(1)}^{jl} = -e^{i\lambda \xi_I} \frac{1}{\lambda} \partial_j [q^{jl}(\nabla \xi)[u^l]]. \tag{483}$$

The derivative ∂_j in (483) will turn out to cost a factor at most ϵ_v^{-1} (see Section (26)). On the other hand, the parametrix gains a factor of λ^{-1}. The way we have chosen ϵ_v ensures us that each iteration of the parametrix gains a factor

$$\frac{\epsilon_v^{-1}}{\lambda} \leq C \frac{\Xi N^{1/L}}{B_\lambda N \Xi} \tag{484}$$

$$\leq \frac{C}{B_\lambda N^{(1-1/L)}} \tag{485}$$

$$\leq \frac{C}{B_\lambda N^{1/2}} \tag{486}$$

since $L \geq 2$.

In the following section, we derive a transport equation for v_ϵ, and derive some estimates that will be necessary for the proof. At this point, we abbreviate η_ϵ for η_{ϵ_v}.

16 The Coarse Scale Flow and Commutator Estimates

In contrast to the argument in [DLS14], we do not mollify the velocity field in the time variable. Instead, by using the fact that v obeys the Euler-Reynolds equation

$$(\partial_t + v^a \partial_a) v^j = \partial^j p + \partial_i R^{ij} \tag{487}$$

we derive a transport equation for v_ϵ^j as follows.

By convolving (487) with $\eta_{\epsilon+\epsilon}$, we can see that $v_\epsilon^j = \eta_\epsilon * \eta_\epsilon * v^j$ also obeys its own transport type equation

$$(\partial_t + v_\epsilon^a \partial_a) v_\epsilon^j = \eta_{\epsilon+\epsilon} * (\partial^j p + \partial_i R^{ij}) + Q^j(v, v) \tag{488}$$

$$= f_\epsilon^j \tag{489}$$

$$Q^j(v, v) = v_\epsilon^a \partial_a v_\epsilon^j - \eta_{\epsilon+\epsilon} * (v^a \partial_a v^j). \tag{490}$$

The quadratic term arises from the failure of the nonlinearity to commute with the averaging, just as the Reynolds stress arises in Section (1). In what follows we will derive estimates for the commutator $Q^j(v,v)$ which should be compared to the commutator estimates used in the proof of energy conservation in [CET94] and the similar commutator estimates used in [CDLS12b]; we give some remarks about the comparison at the end of the section.

According to the Definition (10.1), we expect that the cost of the derivative $(\partial_t + v^a_\epsilon \partial_a)$ should be a factor of $\Xi e_v^{1/2}$, so we expect an estimate of the type $|(\partial_t + v^a_\epsilon \partial_a)v^j_\epsilon| \lesssim \Xi e_v$. However, without observing the cancellation between the two parts of the quadratic term, the only estimate we could expect for $Q^j(v,v)$ would take the form[15] $|Q^j(v,v)| \lesssim \Xi e_v^{1/2}$, and even this estimate would require control over $\|v\|_{C^0}$ which we have not assumed.

The key to observing cancellation in the quadratic term is to use the control over the higher-frequency part of v, and obtain cancellation from the lower-frequency parts. We accomplish this task through the following commutator calculation:

$$Q^j(v,v) = v^a_\epsilon \partial_a v^j_\epsilon - \eta_{\epsilon+c} * (v^a \partial_a v^j) \tag{491}$$

$$= v^a_\epsilon \partial_a(\eta_\epsilon * \eta_\epsilon * v^j) - \eta_\epsilon * \eta_\epsilon * (v^a \partial_a v^j) \tag{492}$$

$$- \eta_\epsilon * (v^a_\epsilon \partial_a(\eta_\epsilon * v^j)) + [v^a_\epsilon \partial_a, \eta_\epsilon *](\eta_\epsilon * v^j)$$
$$- \eta_\epsilon * (v^a \partial_a(\eta_\epsilon * v^j)) + \eta_\epsilon * ([v^a \partial_a, \eta_\epsilon *]v^j) \tag{493}$$

where we can express the commutator terms as

$$[v^a_\epsilon \partial_a, \eta_\epsilon *](\eta_\epsilon * v^j) = \int_{\mathbb{R}^3} (v^a_\epsilon(x-y) - v^a_\epsilon(x))\partial_a(\eta_\epsilon * v^j)(x-y)\eta_\epsilon(y)dy \tag{494}$$

$$= \int_0^1 \int_{\mathbb{R}^3} \partial_i v^a_\epsilon(x-sy)\partial_a(\eta_\epsilon * v^j)(x-y)y^i\eta_\epsilon(y)dyds \tag{495}$$

$$= \epsilon_v \int_0^1 \int_{\mathbb{R}^3} \partial_i v^a_\epsilon(x-sy)\partial_a(\eta_\epsilon * v^j)(x-y)\tilde{\eta}^i_\epsilon(y)dyds \tag{496}$$

and similarly

$$\eta_\epsilon * ([v^a \partial_a, \eta_\epsilon *]v^j)(x)$$
$$= \epsilon_v \int_0^1 \eta_\epsilon * \left[\int_{\mathbb{R}^3} \partial_i v^a(x-sy_1)\partial_a v^j(x-y_1)\tilde{\eta}^i_\epsilon(y_1)dy_1 \right] ds \tag{497}$$

$$= \epsilon_v \int_0^1 \int_{\mathbb{R}^3 \times \mathbb{R}^3} \partial_i v^a(x-y_2-sy_1)\partial_a v^j(x-y_2-y_1)\tilde{\eta}^i_\epsilon(y_1)dy_1 \eta_\epsilon(y_2)dy_2 ds. \tag{498}$$

---

[15] For example, in $v^a_\epsilon \partial_a v^j_\epsilon$, v^a_ϵ is bounded in size by a constant and $\partial_a v^j_\epsilon$ is bounded by $\Xi e_v^{1/2}$.

It is now clear that the commutator terms can be estimated using our control of only the derivatives of v. A similar expression for the remaining terms allows us to finally observe the cancellation between the low frequency contributions to Q

$$\eta_\epsilon * (v_\epsilon^a \partial_a(\eta_\epsilon * v^j)) - \eta_\epsilon * (v^a \partial_a(\eta_\epsilon * v^j)) = \eta_\epsilon * [(v_\epsilon^a - v^a)\partial_a(\eta_\epsilon * v^j)]$$

$$= \epsilon_v \int_0^1 \eta_\epsilon * \left[\int_{\mathbb{R}^3} \partial_i v^j(x - sy)\partial_a(\eta_\epsilon * v^j)(x-y)\tilde{\eta}_\epsilon^i(y)dy \right] ds. \quad (499)$$

Each of the terms (496), (498) and (499) leads to a bound

$$\|Q\|_{C^0} \leq C\epsilon_v \Xi^2 e_v \quad (500)$$

$$\leq CN^{-1/L}\Xi e_v. \quad (501)$$

Differentiating (496), (498) and (499) up to order $L - 1$ costs a factor of Ξ according to the bounds (241) and (474), and beyond order $L - 1$, each derivative costs $N^{1/L}\Xi$ as it hits the mollifiers or as we apply the bounds in (474). These observations lead to the estimates

$$\|\nabla^k Q\|_{C^0} \leq N^{[(k+1-L)_+ - 1]/L}\Xi^{k+1} e_v \quad (502)$$

for the commutator.

These bounds are better than the bounds for

$$\|\nabla^k \eta_{\epsilon+\epsilon} * \nabla p\|_{C^0} \leq N^{[(k+1-L)_+]/L}\Xi^{k+1} e_v \quad (503)$$

coming from (242) by a factor of $N^{-1/L}$. The bounds on

$$\|\nabla^k \eta_{\epsilon+\epsilon} * \nabla R\|_{C^0} \leq N^{[(k+1-L)_+]/L}\Xi^{k+1} e_R \quad (504)$$

coming from (243) are also better than those for the pressure because $e_R \leq e_v$.

For v_ϵ^j defined as above, we therefore have the following bounds.

Theorem 16.1 (Coarse Scale Force Estimates).

$$\|\nabla^\alpha (\partial_t + v_\epsilon^j \partial_j)v_\epsilon^j\|_{C^0} \leq C_\alpha N^{(1+|\alpha|-L)_+/L}\Xi^{1+|\alpha|} e_v \quad (505)$$

for all $|\alpha| \geq 0$.[16]

---

[16] As the above analysis reveals, the double mollification of Section (15) is useful because it allows us to easily prove an estimate for derivatives of order higher than L. Another way to make sure these estimates are available is to choose the mollifier to be a projection $P_{\leq k}$ to frequencies of wavelength 2^{-k} less than ϵ. With such a choice of mollifier, it is then easy to see that the quadratic term $Q(v, v)$ has bounded support in frequency space, and therefore obeys estimates for its higher derivatives as well. Thus, the use of the double mollification plays the same role in the analysis as the formal identity "$P_{\leq k}^2 = P_{\leq k}$" in the standard Littlewood-Paley theory.

In particular, for $|\alpha| = 0$,

$$\|f_\epsilon^j\|_{C^0} = \|(\partial_t + v_\epsilon^a \partial_a)v_\epsilon^j\|_{C^0} \leq C\Xi e_v. \tag{506}$$

Equation (488) has a physical interpretation in terms of the coarse scale flow, which we define here for future reference.

Definition 16.1. Let $\Phi_s(t, x) = (\Phi^0, \Phi^i) : \mathbb{R} \times \mathbb{R} \times \mathbb{T}^3 \to \mathbb{R} \times \mathbb{T}^3$ be the unique solution to the ODE

$$\frac{d}{ds}\Phi_s^0(t, x) = 1 \tag{507}$$

$$\frac{d}{ds}\Phi_s^j(t, x) = v_\epsilon^j(\Phi_s(t, x)) \qquad j = 1, 2, 3 \tag{508}$$

$$\Phi_0(t, x) = (t, x). \tag{509}$$

We call $\Phi_s(t, x) = (t + s, \Phi_s^j(t, x))$ the **coarse scale flow**.

According to Equation (488), any particle which travels along the coarse scale flow experiences an acceleration

$$\frac{d^2}{ds^2}\Phi_s^j(t, x) = (\partial_t + v_\epsilon^a \partial_a)v^j(\Phi_s(t, x)) \tag{510}$$

$$= f_\epsilon^j(\Phi_s(t, x)) \tag{511}$$

where the acceleration f_ϵ^j is given by the right-hand side of (488) and is bounded by $|f_\epsilon^j| \leq C\Xi e_v$.

The method used here to obtain commutator estimates differs from the computations in [CET94] and [CDLS12b] in that our proof only involves commuting mollifiers with differential operators $v^a \partial_a$ and involves a double mollification trick in order to bound higher order derivatives. On the other hand, since both v^a and v_ϵ^a are divergence free, the commutator can be expressed in the form

$$Q^j(v, v) = \partial_a[v_\epsilon^a v_\epsilon^j - \eta_{\epsilon+\epsilon} * (v^a v^j)]$$

as in [CET94] and [CDLS12b] so in particular the estimates (502) we have obtained for spatial derivatives of Q^j can be compared to those stated in Lemma 2.1 of [CDLS12b] for commutators of the above form, which are similar in nature although they are not strong enough for the bounds (502).

17 Transport Estimates

In this section, we use the estimates of Section (16) to derive estimates for quantities which are transported by the coarse scale flow and for their derivatives.

17.1 Stability of the Phase Functions

We start with the phase functions ξ_I which satisfy the transport equation

$$(\partial_t + v_\epsilon^j \partial_j)\xi_I = 0 \tag{512}$$

$$\xi_I(t(I), x) = \hat{\xi}_I \tag{513}$$

with initial data $\hat{\xi}_I$ at time $t(I)$ as described in Sections (7.1) and (7.3).

Our first objective is to choose the lifespan parameter τ sufficiently small so that all the phase functions which appear in the analysis can be guaranteed to remain nonstationary in the time interval $|t - t(I)| \leq \tau$, and so that the stress equation studied in Section (7.3) can be solved. In order for these requirements to be met, we will choose τ small enough so that the gradients of the phase functions do not depart dramatically from their initial configurations.

The **lifespan parameter** τ will be of the form

$$\tau = b\Xi^{-1}e_v^{-1/2} \tag{514}$$

$$b \leq 1 \tag{515}$$

where b is a dimensionless parameter that must be sufficiently small.

The following proposition bounds the separation of the phase gradients from their initial values in terms of b.

Proposition 17.1. *Let Φ_s be the coarse scale flow defined by (16.1) and let $\hat{\xi}_I(x)$ be the initial conditions defined in Section (7.3.3).*

There exists a constant A such that for all $|s| < b\Xi^{-1}e_v^{-1/2}$, and all $I \in \mathcal{I}$

$$|\nabla\xi_I(\Phi_s(t(I), x)) - \nabla\hat{\xi}_I(x)| \leq Ab. \tag{516}$$

As a consequence,

$$\| |\nabla\xi_I(t, x)) - \nabla\hat{\xi}_I| \,\|_{C^0} \leq Ab \tag{517}$$

for all t in the interval $|t - t(I)| \leq b\Xi e_v^{-1/2}$.

With this proposition in hand, we will require the following.

Requirement 17.1.

$$b \trianglelefteq b_0 \tag{518}$$

where $b_0 < 1$ is a constant ensuring that

$$\| |\nabla\xi_I|^{-1} \|_{C^0} \leq 3|\nabla\hat{\xi}_I|^{-1} \tag{519}$$

$$\| |\nabla(\xi_I + \xi_J)|^{-1} \|_{C^0} \leq 3|\nabla(\hat{\xi}_I + \hat{\xi}_J)|^{-1} \qquad J \neq \bar{I} \tag{520}$$

on $|t - t(I)| \leq b_0 \Xi^{-1} e_v^{-1/2}$, and also ensuring that

$$\| \, |\nabla \xi_I(t, x)) - \nabla \hat{\xi}_I(x)| \, \|_{C^0} \leq c \tag{521}$$

where c guarantees the conditions in Proposition (7.1) are both satisfied. Proposition (17.1) is a consequence of the inequality.

Proposition 17.2. *There exist absolute constants C_1, C_2 such that for all $|s| \leq b \Xi^{-1} e_v^{-1/2}$ we have*

$$|\nabla \xi_I(\Phi_s(t_I, x)) - \nabla \hat{\xi}_I(x)| \leq C_1 e^{C_2 \Xi e_v^{1/2} |s|} b \tag{522}$$

$$\leq C_1 e^{C_2 b} b. \tag{523}$$

Proof. We study the evolution of the quantity

$$u(s) = |\nabla \xi_I(\Phi_s(t(I), x)) - \nabla \hat{\xi}_I(x)|^2 \tag{524}$$

as the flow parameter s varies. To prepare to compute $\frac{du}{ds}$ we first differentiate the transport equation

$$(\partial_t + v_\epsilon^j \partial_j) \xi_I = 0 \tag{525}$$

to obtain the equation

$$(\partial_t + v_\epsilon^j \partial_j) \partial^l \xi_I = -\partial^l v_\epsilon^j \partial_j \xi_I \tag{526}$$

governing the evolution of $\nabla \xi_I$. Using this law, we compute

$$\frac{d}{ds} |\nabla \xi_I(\Phi_s(t_I, x)) - \nabla \hat{\xi}_I(x)|^2 \tag{527}$$

$$= 2(\partial_t + v_\epsilon^j(\Phi_s) \partial_j) \partial^l \xi_I(\Phi_s)(\partial_l \xi_I(\Phi_s) - \partial_l \hat{\xi}_I)$$

$$= -2 \partial^l v_\epsilon^j \partial_j \xi_I(\Phi_s)(\partial_l \xi_I(\Phi_s) - \partial_l \hat{\xi}_I) \tag{528}$$

$$= -2 \partial^l v_\epsilon^j (\partial_j \xi_I(\Phi_s) - \partial_j \hat{\xi}_I)(\partial_l \xi_I(\Phi_s) - \partial_l \hat{\xi}_I) \tag{529}$$

$$- 2 \partial^l v_\epsilon^j \partial_j \hat{\xi}_I (\partial_l \xi_I(\Phi_s) - \partial_l \hat{\xi}_I)$$

which gives an estimate of the form

$$|\frac{du}{ds}| \leq C_2 \Xi e_v^{1/2} u + 2 C_1 \Xi e_v^{1/2} u^{1/2}. \tag{530}$$

In terms of the dimensionless variable $\bar{t} = \Xi e_v^{1/2} s$, the same inequality can be stated

$$|\frac{du}{d\bar{t}}| \leq C_2 u + 2 C_1 u^{1/2}. \tag{531}$$

107

By the time reversal symmetry $\bar{t} \mapsto -\bar{t}$, it suffices to prove (522) for $\bar{t} > 0$. In this case, we achieve the bound by first using (531) to deduce

$$\frac{du}{d\bar{t}} - C_2 u \leq 2C_1 u^{1/2} \tag{532}$$

$$e^{-C_2\bar{t}}\frac{du}{d\bar{t}} - C_2 e^{-C_2\bar{t}} u \leq 2C_1 e^{-C_2\bar{t}} u^{1/2} \tag{533}$$

$$\frac{d}{d\bar{t}}(e^{-C_2\bar{t}} u) \leq 2C_1 e^{-C_2\bar{t}} u^{1/2}. \tag{534}$$

In terms of the variable $U = e^{-C_2\bar{t}} u$, inequality (534) takes the form

$$\frac{dU}{d\bar{t}} \leq 2C_1 e^{-\frac{C_2\bar{t}}{2}} U^{1/2}. \tag{535}$$

To deduce a bound for v, we can integrate the above inequality by observing that for any $\epsilon > 0$

$$\frac{d(U+\epsilon)}{d\bar{t}} \leq 2C_1 e^{-\frac{C_2\bar{t}}{2}} (U+\epsilon)^{1/2} \tag{536}$$

$$\frac{1}{2(U+\epsilon)^{1/2}}\frac{d(U+\epsilon)}{d\bar{t}} = \frac{d(U+\epsilon)^{1/2}}{d\bar{t}} \leq C_1 e^{-\frac{C_2\bar{t}}{2}}. \tag{537}$$

Integrating from $\bar{t} = 0$ to $\bar{t} = \bar{s}$ for $\bar{s} = s\Xi e_v^{1/2}$ in the range $0 \leq \bar{s} \leq b$ and taking $\epsilon \to 0$, we obtain

$$U^{1/2}(\bar{s}) - U^{1/2}(0) \leq C_1 \int_0^{\bar{s}} e^{-\frac{C_2\bar{t}}{2}} d\bar{t} \tag{538}$$

$$\leq C_1 \bar{s} \tag{539}$$

$$\leq C_1 b. \tag{540}$$

In our case, $U(0) = 0$ because $\nabla\hat{\xi}_I(x)$ is the initial value for the phase gradient $\nabla\ddot{\xi}_I(\Phi_s(t(I), x))$. Thus, returning to the initial variable u,

$$e^{-\frac{C_2}{2}\bar{s}} u^{1/2}(\bar{s}) \leq C_1 b \tag{541}$$

$$u^{1/2}(\bar{s}) \leq C_1 e^{\frac{C_2\bar{s}}{2}} b \tag{542}$$

which is an inequality of the form (522) with a different value for C_2 in terms of the original parameter $s = \Xi^{-1} e_v^{-1/2} \bar{s}$. $\qquad\square$

17.2 Relative Velocity Estimates

We now collect estimates for the derivatives of the phase functions. In terms of the foliation generated by the level sets of the phase functions ξ_I, estimates for derivatives $\nabla^k \xi_I$ encode quantitative information about the density and regularity of the foliation. Each leaf $\xi_I = C$ moves with the coarse scale flow, so changes in the regularity of the foliation arise from variations of the velocity field $v_\epsilon(t, x)$ in space. Hence we refer to these estimates collectively as relative velocity estimates. The intuitive considerations above are captured by the structure of the transport equations in the analysis. For example, the bounds we obtain never depend on the C^0 norm of v_ϵ, which is consistent with the Galilean invariance of the equation.

We now generalize the argument of Section (17.1) to obtain bounds for the higher derivatives $\nabla^k \xi_I$. From the estimates on the coarse scale velocity field $\|\nabla^a v_\epsilon\|_{C^0} \leq \Xi^{|a|} N^{(a-L)_+/L} e_v^{1/2}$, we see that each derivative of the velocity field v_ϵ costs a factor of Ξ, with an extra cost of $N^{1/L}$ for each derivative beyond order L. We expect the same cost in the estimates for the derivatives of ξ, so our goal is to prove the following

Proposition 17.3. *There exist constants C_m depending on m such that for all multi-indices γ of order $|\gamma| = m$ we have*

$$\|\nabla \xi_I\|_{C^0} \leq C \tag{543}$$

$$\|\nabla^\gamma \xi_I\|_{C^0} \leq C_{|\gamma|} \Xi^{|\gamma|-1} N^{(|\gamma|-L)_+/L} \qquad \text{for all } |\gamma| \geq 1. \tag{544}$$

Motivated by the statement of Proposition (17.3) and the structure of the transport equations to follow, we introduce a weighted energy to measure the pointwise values of the M-jet of ξ_I.

Definition 17.1. Let $\xi : \mathbb{T}^3 \times \mathbb{R} \to \mathbb{R}$ be a real valued function, and let Ξ, e_v, N and L be as in (10.1) and (10.1). For $1 \leq M$, we define the M'th order dimensionless energy of ξ to be

$$E_M[\xi](t, x) = \sum_{m=1}^{M} \sum_{|\gamma|=m} \Xi^{-2(m-1)} N^{-2(m-L)_+/L} (\partial^\gamma \xi)^2 \tag{545}$$

where the summation runs over all multi-indices γ with $1 \leq |\gamma| \leq M$.

In terms of the weighted norm in Definition (17.1), the bound (17.3) can be deduced from the following more general theorem, which will be used again later on.

Proposition 17.4. *Suppose that ξ satisfies the transport equation*

$$(\partial_t + v_\epsilon^j \partial_j)\xi = 0. \tag{546}$$

Then there exists a constant $C = C_M$ such that for all $s \in \mathbb{R}$ and all $(t, x) \in \mathbb{R} \times \mathbb{T}^3$

$$E_M[\xi](\Phi_s(t, x)) \leq e^{C\Xi e_v^{1/2}|s|} E_M[\xi](t, x). \tag{547}$$

In particular, for $|s| \leq \tau = b\Xi^{-1} e_v^{-1/2}$ with $b \leq 1$, we have a bound

$$E_M[\xi](\Phi_s(t, x)) \leq C E_M[\xi](t, x) \qquad \text{for all } |s| \leq \tau. \tag{548}$$

Proof. For the case $M = 1$, the weighted norm coincides with the usual absolute value $|\nabla \xi|^2$. To study the evolution of $|\nabla \xi|$ under the coarse scale flow, we use the equation

$$(\partial_t + v_\epsilon^j)\partial^\gamma \xi = -\partial^\gamma v_\epsilon^j \partial_j \xi \tag{549}$$

to obtain a formula involving the deformation tensor of v_ϵ

$$\frac{d}{ds}|\nabla \xi|^2(\Phi_s) = 2 \sum_{|\gamma|=1} (\partial_t + v_\epsilon^j \partial_j)\partial^\gamma \xi \partial_\gamma \xi \tag{550}$$

$$= -2 \sum_{|\gamma|=1} \partial^\gamma v_\epsilon^j \partial_j \xi \partial_\gamma \xi \tag{551}$$

$$= - \sum_{|\gamma|=1} (\partial^\gamma v_\epsilon^j + \partial^j v_\epsilon^\gamma)\partial_j \xi \partial_\gamma \xi. \tag{552}$$

All we need to extract from this evolution formula is the estimate

$$\left| \frac{d}{ds}|\nabla \xi|^2(\Phi_s) \right| \leq C \Xi e_v^{1/2} |\nabla \xi|^2(\Phi_s) \tag{553}$$

from which (547) follows by Gronwall's inequality.

For $M = 2$, we prepare to prove an estimate

$$\left| \frac{d}{ds}E_2[\xi](\Phi_s) \right| \leq C_2 \Xi e_v^{1/2} E_2[\xi](\Phi_s) \tag{554}$$

by first differentiating (549) to obtain

$$(\partial_t + v_\epsilon^j \partial_j)\partial^{\gamma_1}\partial^{\gamma_2}\xi = -\partial^{\gamma_1}[\partial^{\gamma_2}v_\epsilon^j \partial_j \xi] - \partial^{\gamma_1}v_\epsilon^j \partial_j \partial^{\gamma_2}\xi \tag{555}$$

$$= -(\partial^{\gamma_2}v_\epsilon^j \partial_j \partial^{\gamma_1}\xi + \partial^{\gamma_1}v_\epsilon^j \partial_j \partial^{\gamma_2}\xi) - \partial^{\gamma_1}\partial^{\gamma_2}v_\epsilon^j \partial_j \xi. \tag{556}$$

We can use this equation to obtain (554) by first observing that

$$E_2[\xi] = \sum_{|\gamma|=2} \Xi^{-2}\left(\sum_{|\gamma|=2} |\partial^\gamma \xi|^2 \right) + |\nabla \xi|^2 \tag{557}$$

so that (554) follows from an inequality

$$\Xi^{-2}\left| \frac{d}{ds}|\partial^\gamma \xi|^2(\Phi_s) \right| \leq C_\gamma \Xi e_v^{1/2} E_2[\xi] \qquad \text{for all } |\gamma| = 2. \tag{558}$$

The above bound on

$$\frac{d}{ds}|\partial^\gamma \xi|^2(\Phi_s) = 2[(\partial_t + v_\epsilon^j \partial_j)\partial^{\gamma_1}\partial_{\gamma_2}\xi]\partial_{\gamma_1}\partial_{\gamma_2}\xi \tag{559}$$

follows from multiplying Equation (555) by $\Xi^{-2}\partial_{\gamma_1}\partial_{\gamma_2}\xi$, and observing that

$$\Xi^{-2}\partial_{\gamma_1}\partial_{\gamma_2}\xi(\partial^{\gamma_1}\partial^{\gamma_2}v_\epsilon^j\partial_j\xi) = (\Xi^{-1}\partial^{\gamma_1}\partial^{\gamma_2}v_\epsilon^j)(\Xi^{-1}\partial_{\gamma_1}\partial_{\gamma_2}\xi)(\partial_j\xi) \qquad (560)$$

$$\leq \Xi e_v^{1/2} \; E_2^{1/2}[\xi](\Phi_s) \cdot E_2^{1/2}[\xi](\Phi_s) \qquad (561)$$

$$\leq \Xi e_v^{1/2} \; E_2[\xi](\Phi_s). \qquad (562)$$

Clearly, the terms arising from the remaining part in (555) obey the same estimate.

In general, we can prove the differential inequality

$$\left| \frac{d}{ds} E_M[\xi](\Phi_s) \right| \leq C_M \Xi e_v^{1/2} E_M[\xi](\Phi_s(t,x)) \qquad (563)$$

inductively by showing that for all $|\gamma| = M$ we have a bound

$$\Xi^{-2(M-1)} N^{-2(M-L)_+/L} \left| \frac{d}{ds} |\partial^\gamma \xi|^2(\Phi_s) \right| \leq C_\gamma \Xi e_v^{1/2} E_M[\xi](\Phi_s). \qquad (564)$$

The transport equation obeyed by a higher order derivative $\partial^\gamma \xi = \partial^{\gamma_1} \cdots \partial^{\gamma_M} \xi$ is of the form

$$(\partial_t + v_\epsilon^j \partial_j)\partial^\gamma \xi = \sum_{\substack{\alpha+\beta=\gamma \\ |\alpha|\geq 1}} c_{\alpha,\beta}^\gamma \partial^\alpha v_\epsilon^j \partial_j(\partial^\beta \xi). \qquad (565)$$

After multiplying this equation by $\Xi^{-2(M-1)} N^{-2(M-L)_+/L}\partial_\gamma\xi$, each term has the form

$$\Xi^{-2(M-1)} N^{-2(M-L)_+/L}\partial^\alpha v_\epsilon^j \partial_j(\partial^\beta\xi)\partial_\gamma\xi = P_{(I)} \cdot P_{(II)} \qquad (566)$$

$$P_{(I)} = \Xi^{-(M-1)} N^{-(M-L)_+/L}\partial^\alpha v_\epsilon^j \partial_j(\partial^\beta\xi)$$

$$P_{(II)} = \Xi^{-(M-1)} N^{-(M-L)_+/L}\partial_\gamma\xi$$

$$|\alpha| + |\beta| = M, \quad |\beta| \leq M - 1.$$

In order to prepare to bound these terms in terms of the dimensionless energy, we use the simple observation that for all $\alpha \geq 1, L \geq 1$ and $\beta \geq 0$, we have

$$(\alpha + \beta - L)_+ \geq ((\alpha - 1) - (L - 1))_+ + (\beta - (L-1))_+$$
$$= (\alpha - L)_+ + (\beta + 1 - L)_+. \qquad (567)$$

This inequality follows from the more general inequality (17.1) stated below.

Using (567), we can then distribute the powers of N and Ξ

$$|P_{(I)}| = \Xi^{-(|\alpha|+|\beta|-1)} N^{-(M-L)_+/L} \left|\partial^\alpha v_\epsilon^j \partial_j(\partial^\beta\xi)\right| \qquad (568)$$

$$\leq \left|\Xi^{-|\alpha|+1} N^{-(|\alpha|-L)_+/L}\partial^\alpha v_\epsilon^j\right| \; \left|\Xi^{-|\beta|} N^{-(|\beta|+1-L)_+/L}\partial_j(\partial^\beta\xi)\right|. \qquad (569)$$

Using the estimates

$$\|\partial^\alpha v_\epsilon^j\|_{C^0} \leq \Xi^{|a|} N^{(|a|-L)_+/L} e_v^{1/2} \tag{570}$$

$$\Xi^{-(|\gamma|-1)} N^{-(|\gamma|-L)_+/L} |\partial_\gamma \xi| \leq E_M^{1/2}[\xi] \tag{571}$$

and (569) we bound

$$|P_{(I)}| \leq (\Xi e_v^{1/2}) \cdot E_M^{1/2}[\xi](\Phi_s) \tag{572}$$

$$|P_{(II)}| \leq E_M^{1/2}[\xi](\Phi_s) \tag{573}$$

giving an estimate for (566) of

$$\Xi^{-2(M-1)} N^{-2(M-L)_+/L} |\partial^\alpha v_\epsilon^j \partial_j (\partial^\beta \xi) \partial_\gamma \xi| \leq \Xi e_v^{1/2} E_M[\xi](\Phi_s), \tag{574}$$

which establishes (564), and by Gronwall, we conclude (547). □

To deduce the Proposition (17.3) from Proposition (17.4), we simply observe that the dimensionless norm of ξ_I is initially bounded by

$$E_M[\xi]^{1/2}(t(I), x) = E_M[\hat\xi]^{1/2}(x) \tag{575}$$

$$\leq C_M \tag{576}$$

and that every other point (t, x) on the support of v_I is obtained by flowing for a time less than τ from some other point $(t(I), x')$.

As a corollary to Proposition (548) and Equations (565) for $\xi = \xi_I$, we also have the following bounds.

Proposition 17.5 (Phase-Velocity Estimates). *For all multi-indices γ, k of orders $|\gamma| \geq 0$, and $|k| \geq 0$*

$$\|\nabla^k[(\partial_t + v_\epsilon \cdot \nabla)\nabla^\gamma(\nabla \xi_I)]\|_{C^0} \leq C\Xi^{|k|+|\gamma|+1} e_v^{1/2} N^{(|k|+|\gamma|+1-L)_+/L}. \tag{577}$$

The above bounds are larger than the bounds in (548) by a factor $\Xi e_v^{1/2}$ coming from the material derivative.

The inequality used to count powers of $N^{1/L}$ in the proof of Proposition (17.3) follows from the following general inequality, which will be used similarly in further estimates below.

Lemma 17.1 (The Counting Inequality). *For any non-negative numbers $x_1, x_2, \ldots, x_M \geq 0$ and $Y \geq 0$, we have*

$$(x_1 + x_2 + \ldots + x_M - Y)_+ \geq \sum_i (x_i - Y)_+. \tag{578}$$

Proof. The result for general M follows by induction from the case where $M = 2$, which states that

$$(x_1 - Y)_+ + (x_2 - Y)_+ \leq (x_1 + x_2 - Y)_+. \tag{579}$$

Inequality (579) is obvious when the left-hand side is zero. If exactly one of the two terms, say $(x_1 - Y)_+$, is positive, then the inequality follows from $x_2 \geq 0$. If both $(x_1 - Y)_+$ and $(x_2 - Y)_+$ are positive, then the inequality follows from $Y \geq 0$. $\qquad\square$

17.3   Relative Acceleration Estimates

In this section, we gather estimates for the quantities

$$(\partial_t + v_\epsilon^j \partial_j)^2 \nabla^k \xi_I$$

and their spatial derivatives which will be useful in the analysis to follow.

These estimates along with the phase-velocity estimates of Section (17.2) give quantitative information regarding how the regularity of the foliation $\{\xi_I = C\}$ changes in time from the frame of reference of a particle moving along the coarse scale flow. Since the acceleration in time of these geometric quantities is due to variations in the acceleration that such particles experience, we refer to the collection of estimates that follow as relative acceleration estimates.

In Section (16) we derived the following estimates for the acceleration along the coarse scale flow:

$$\|\nabla^\alpha[(\partial_t + v_\epsilon^j \partial_j)v_\epsilon^l]\|_{C^0} \leq C_\alpha N^{(1+|\alpha|-L)_+/L} \Xi^{1+|\alpha|} e_v \qquad |\alpha| \geq 0 \tag{580}$$

and as in that section, we will abbreviate $f_\epsilon^l = (\partial_t + v_\epsilon^j \partial_j)v_\epsilon^l$.

Starting with

$$(\partial_t + v_\epsilon^b \partial_b)\partial^l \xi_I = -\partial^l v_\epsilon^b \partial_b \xi_I \tag{581}$$

we calculate

$$(\partial_t + v_\epsilon^a \partial_a)[(\partial_t + v_\epsilon^b \partial_b)\partial^l \xi_I] = -(\partial_t + v_\epsilon^a \partial_a)[\partial^l v_\epsilon^b \partial_b \xi_I] \tag{582}$$

$$= -\partial^l v_\epsilon^b(\partial_t + v_\epsilon^a \partial_a)\partial_b \xi_I - [(\partial_t + v_\epsilon^a \partial_a)\partial^l v_\epsilon^b]\partial_b \xi_I \tag{583}$$

$$= \partial^l v_\epsilon^b \partial_b v_\epsilon^a \partial_a \xi_I - [(\partial_t + v_\epsilon^a \partial_a)\partial^l v_\epsilon^b]\partial_b \xi_I \tag{584}$$

and substituting

$$(\partial_t + v_\epsilon^a \partial_a)\partial^l v_\epsilon^b = \partial^l[(\partial_t + v_\epsilon^a \partial_a)v_\epsilon^b] - \partial^l v_\epsilon^a \partial_a v_\epsilon^b \tag{585}$$

$$= \partial^l[f_\epsilon^b] - \partial^l v_\epsilon^a \partial_a v_\epsilon^b \tag{586}$$

leaves

$$(\partial_t + v_\epsilon^a \partial_a)[(\partial_t + v_\epsilon^b \partial_b)\partial^l \xi_I] = -\partial^l f_\epsilon^b \partial_b \xi_I + 2\partial^l v_\epsilon^b \partial_b v_\epsilon^a \partial_a \xi_I. \tag{587}$$

Applying the estimates (580), (474) and Proposition (17.3) to Equation (587) we obtain the following.

113

Proposition 17.6 (Phase-Acceleration Estimates). *For any multi-index α of order $|\alpha| \geq 0$*

$$\|\nabla^\alpha[(\partial_t + v_\epsilon \cdot \nabla)^2 \nabla \xi_I]\|_{C^0} \leq C_{|\alpha|} N^{(2+|\alpha|-L)_+/L} \Xi^{(2+|\alpha|)} e_v. \qquad (588)$$

We can obtain acceleration estimates for higher order derivatives $\nabla^\gamma \xi_I$ by repeatedly commuting spatial derivatives with $(\partial_t + v_\epsilon \cdot \nabla)$ and combining the estimates (588) and (17.3). The resulting bound is of the same form.

Proposition 17.7 (Higher Order Phase-Acceleration Estimates). *For any multi-indices α, γ of orders $|\alpha| \geq 0$ and $|\gamma| \geq 0$*

$$\|\nabla^\alpha[(\partial_t + v_\epsilon \cdot \nabla)^2 \nabla^\gamma (\nabla \xi_I)]\|_{C^0}$$
$$\leq C_{|\alpha|+|\gamma|} N^{(1+|\alpha|+|\gamma|-L)_+/L} \Xi^{(2+|\alpha|+|\gamma|)} e_v \qquad (589)$$

One can summarize the estimates (17.3), (577), (589) succinctly in the form

$$\|\nabla^\alpha[(\partial_t + v_\epsilon \cdot \nabla)^r \nabla^\gamma (\nabla \xi_I)]\|_{C^0}$$
$$\leq C N^{((r-1)_+ + |\gamma| + |\alpha| - L)_+/L} \Xi^{r+|\alpha|+|\gamma|} e_v^{r/2} \qquad (590)$$

and more generally we can commute the spatial and material derivatives to obtain the following.

Proposition 17.8 (General Phase-Acceleration Estimates 1).

$$\|\nabla^\alpha (\partial_t + v_\epsilon \cdot \nabla) \nabla^\beta (\partial_t + v_\epsilon \cdot \nabla) \nabla^\gamma (\nabla \xi_I)\|_{C^0}$$
$$\leq C N^{(1+|\alpha|+|\beta|+|\gamma|-L)_+/L} \Xi^{2+|\alpha|+|\gamma|} e_v$$

We summarize some important points from the analysis.

- The gradients $\nabla \xi_I$ are bounded in size.

- Each spatial derivative of $\nabla \xi$ costs a factor of Ξ, which is consistent with dimensional analysis.

- Each material derivative $(\partial_t + v_\epsilon \cdot \nabla)^r$, $|r| \leq 2$, costs a factor of $\Xi e_v^{1/2}$, which is also consistent with dimensional analysis.

- After taking L derivatives of ξ, the estimates also lose a factor of $a_v^{-1} N^{1/L}$ for each additional derivative taken (although the first material derivative does not count in this regard).

- **In particular**, since $L \geq 2$, more than two derivatives of ξ must be taken before this extra loss appears.

114

Let us introduce some notation to make the estimates more readable. Denote by $D^{(k,r)}$ any differential operator of the form

$$D^{(k,r)} = \nabla^{a_1}(\partial_t + v_\epsilon \cdot \nabla)^{r_1}\nabla^{a_2}(\partial_t + v_\epsilon \cdot \nabla)^{r_2}\nabla^{a_3} \qquad (591)$$

$$k = a_1 + a_2 + a_3 \qquad (592)$$

$$r = r_1 + r_2 \qquad (593)$$

$$a_i, r_i \geq 0 \qquad (594)$$

$$r \leq 2. \qquad (595)$$

The general phase-acceleration estimate can be written concisely as the following.

Proposition 17.9 (General Phase-Acceleration Estimates 2).

$$\|D^{(k,r)}\nabla\xi_I\|_{C^0} \leq C_{k+r} N^{((r-1)_+ + k + 1 - L)_+ / L}\Xi^{k+r} e_v^{r/2} \qquad (596)$$

18 Mollification along the Coarse Scale Flow

With the transport estimates of Section (17) in hand, we are ready to discuss how to construct the appropriate mollification of the Reynolds stress.

18.1 The Problem of Mollifying the Stress in Time

Unlike the velocity field, which was only mollified in the spatial variables and which earned its time-regularity through the Euler-Reynolds equation, the Reynolds stress must be mollified in both space and time. To see that a mollification is necessary, first observe that according to the construction of $v_I \approx R^{1/2}$ in Section (7.3), the Transport term

$$\partial_j Q_T^{jl} \approx \sum_I e^{i\lambda\xi_I}(\partial_t + v_\epsilon^j \partial_j)v_I^l \qquad (597)$$

involves a time derivative of R and we cannot obtain C^0 bounds for the stress Q_T arising from this term without having control over $(\partial_t + v^j\partial_j)R$. On the other hand, in order to proceed to the next stage of the iteration, we must show that $(\partial_t + v_I^a\partial_a)Q_T^{jl}$ obeys good bounds as well, and verifying these bounds will require us to take a second material derivative of R according to Equation (597).

According to the above considerations, it is necessary to mollify R in time. It is also necessary to mollify along the flow lines of v, rather than in the t direction, because R fluctuates too rapidly in the t direction, whereas the material derivative $(\partial_t + v^j\partial_j)R$ obeys better bounds. Unlike mollification along the t direction, mollification along the flow is also consistent with the Galilean invariance of the equations.

115

18.2 Mollifying the Stress in Space and Time

After constructing the mollification R_ϵ of R, we will need to have estimates for the quantities:

- $\|R_\epsilon\|_{C^0}$

- $\|\nabla^k R_\epsilon\|_{C^0}$, $|k| \geq 0$

- $\|(\partial_t + v_\epsilon \cdot \nabla)^r R_\epsilon\|_{C^0}$, $r = 1, 2$

- $\|\nabla^k(\partial_t + v_\epsilon \cdot \nabla)^r R_\epsilon\|_{C^0}$, $r = 1, 2$ and $|k| \geq 0$

With these demands in mind, we construct the mollification in two steps. First we average in space by defining

$$R_{\epsilon_x}(t, x) = \eta_{\epsilon_x} * \eta_{\epsilon_x} * R(t, x) \tag{598}$$

$$= \int \int R(t, x + y_1 + y_2) \eta_{\epsilon_x}(y_1) \eta_{\epsilon_x}(y_2) dy_1 dy_2. \tag{599}$$

The double mollification here plays the same role as the double mollification in the construction of v_ϵ in Section (15).

We then use the coarse scale flow Φ defined in (16.1) to average in time by defining

$$R_\epsilon(t, x) = R_{\epsilon_t \epsilon_x}(t, x) = \eta_{\epsilon_t} *_\Phi R_{\epsilon_x}(t, x) \tag{600}$$

$$= \int R_{\epsilon_x}(\Phi_s(t, x)) \eta_{\epsilon_t}(s) ds. \tag{601}$$

For the purpose of our analysis here, it is not important that we have chosen to mollify first in space and then in time rather than in the opposite order, because we will choose ϵ_t and ϵ_x small enough so that these operations commute up to acceptable errors. In general, though, if $\epsilon_t = T$ is too large, the time T flow of an ϵ_x-ball can be deformed around $\mathbb{T}^3$ in a complicated way, and will look very different from the ϵ_x neighborhood of a single trajectory of time duration T.

18.3 Choosing Mollification Parameters

As a goal, we will require that the error term generated by this mollification constitutes a small fraction of the allowable stress

$$\|R - R_\epsilon\|_{C^0} \trianglelefteq \frac{e_v^{1/2} e_R^{1/2}}{100N}. \tag{602}$$

During this mollification $R \to R_\epsilon$, we are also not allowed to enlarge the support of R by an amount larger than $\Xi^{-1} e_v^{-1/2}$, as we have only assumed a lower bound of

$$e(t) \geq K e_R, \qquad t \in [a - \Xi^{-1} e_v^{-1/2}, b + \Xi e_v^{1/2}] \tag{603}$$

$$\text{supp } R \subseteq [a, b] \times \mathbb{T}^3 \tag{604}$$

and this lower bound is necessary in order to solve the stress equation in Section (7.3.6).

In pursuit of the goal (602), we define the time-averaged stress

$$R_{\epsilon_t}(t, x) = \eta_{\epsilon_t} *_\Phi R(t, x) \tag{605}$$

$$= \int R(\Phi_s(t, x))\eta_{\epsilon_t}(s)ds \tag{606}$$

and decompose the error into two parts

$$R - R_\epsilon = (R - R_{\epsilon_t}) + (R_{\epsilon_t} - R_{\epsilon_t \epsilon_x}) \tag{607}$$

$$= (R - R_{\epsilon_t}) + \eta_{\epsilon_t} *_\Phi (R - R_{\epsilon_x}). \tag{608}$$

This particular decomposition of the error is sensible because the tensor $R_\epsilon(t, x)$ is obtained by averaging R over an ϵ_x-neighborhood of the time ϵ_t-flow from the point (t, x).

We first discuss the second of these error terms. It is clear that

$$\|\eta_{\epsilon_t} *_\Phi (R - R_{\epsilon_x})\|_{C^0} \le \|R - R_{\epsilon_x}\|_{C^0} \tag{609}$$

so in pursuit of the goal (602), we will set ϵ_x to achieve the goal

$$\|R - R_{\epsilon_x}\|_{C^0} \trianglelefteq \frac{e_v^{1/2} e_R^{1/2}}{200N}. \tag{610}$$

From Section (14) we have the estimate

$$\|R - R_{\epsilon_x}\|_{C^0} \le \|R - \eta_{\epsilon_x} * R\|_{C^0} + \|\eta_{\epsilon_x} * (R - \eta_{\epsilon_x} * R)\|_{C^0} \tag{611}$$

$$\le C\epsilon_x^L \Xi^L e_R. \tag{612}$$

Since $e_v \ge e_R$, we can achieve the goal (610) by choosing

$$\epsilon_x = a_R \Xi^{-1} N^{-1/L} \tag{613}$$

for the appropriately chosen small constant $a_R > 0$. Note that this choice coincides with the scale for the spatial mollification of v in Section (15).

In order to achieve the goal (602), it now suffices to pick ϵ_t small enough so that

$$\|R - R_{\epsilon_t}\|_{C^0} \trianglelefteq \frac{e_v^{1/2} e_R^{1/2}}{200N}. \tag{614}$$

In pursuit of (614), we calculate

$$R(t, x) - R_{\epsilon_t}(t, x) = \int (R(t, x) - R(\Phi_s(t, x)))\eta_{\epsilon_t}(s)ds \tag{615}$$

$$= -\int \left[\int_0^1 \frac{d}{du} R(\Phi_{us}(t, x)) du \right] \eta_{\epsilon_t}(s) ds \tag{616}$$

$$= -\int_0^1 \left[\int [(\partial_t + v_\epsilon^a \partial_a)R](\Phi_{us}(t, x))s\eta_{\epsilon_t}(s)ds \right] du \tag{617}$$

which gives the bound

$$\|R(t,x) - R_{\epsilon_t}(t,x)\|_{C^0} \leq C\epsilon_t \|(\partial_t + v_\epsilon^j \partial_j)R\|_{C^0} \tag{618}$$

$$\leq C\epsilon_t (\|(\partial_t + v^j \partial_j)R\|_{C^0} + \|(v_\epsilon^j - v^j)\partial_j R\|_{C^0}). \tag{619}$$

Applying the bounds (244) and (470), we have an estimate of the form

$$\|R(t,x) - R_{\epsilon_t}(t,x)\|_{C^0} \leq C\epsilon_t \Xi e_v^{1/2} e_R \tag{620}$$

for some absolute constant C, so that we can finally achieve the goal (614) and consequently (602) by setting

$$\epsilon_t \equiv c\Xi^{-1} N^{-1} e_R^{-1/2} \tag{621}$$

for an appropriately small constant $c > 0$.

At this point, we have to check that

$$\epsilon_t \trianglelefteq \Xi^{-1} e_v^{-1/2} \tag{622}$$

for two reasons:

- The flow Φ_s and its derivatives only remain under control for times up to $c\Xi^{-1} e_v^{-1/2}$.

- In order to solve the Stress Equation, Equation (128), we must be sure that the tensor

$$\varepsilon^{jl} = -\frac{\mathring{R}_\epsilon^{jl}}{e(t)}$$

encountered in Section (7.3.6) is bounded by an appropriate constant specified in (147). Since $\|R_\epsilon\|_{C^0} \leq e_R$, the bound (147) is achieved as long as the lower bound

$$e(t) \geq K e_R$$

is satisfied on the support of R_ϵ. In the hypotheses of the Main Lemma (10.1), the interval on which this lower bound is satisfied is assumed to be an amount $\theta = \Xi^{-1} e_v^{-1/2}$ larger than the time interval supporting R itself.

Thankfully, the bound (622) follows from (621) (choosing c smaller if necessary) and the assumption (252) in Lemma (10.1), which implies that

$$N \geq \left(\frac{e_v}{e_R}\right)^{1/2}.$$

18.4 Estimates for the Coarse Scale Flow

In the following section, we will derive bounds for the mollified stress

$$R_\epsilon(t, x) = \int R_{\epsilon_x}(\Phi_s(t, x)) \eta_{\epsilon_t}(s) ds \qquad (623)$$

$$= \int R_{\epsilon_x}(t + s, \Phi_s^i(t, x)) \eta_{\epsilon_t}(s) ds \qquad (624)$$

and its derivatives.

As the formula (623) suggests, it will be important to estimate the extent to which the flow Φ_s deforms the geometry of $\mathbb{R} \times \mathbb{T}^3$ during the time $|s| \leq \epsilon_t \leq c\Xi^{-1} e_v^{-1/2}$. This deformation estimate can be summarized by the following proposition.

Proposition 18.1. *For every multi-index a with $|a| \geq 1$, there exist constants C_1 and C_2 such that the spatial derivative*

$$\partial_a \Phi_s^i(t, x) : \mathbb{R} \times \mathbb{R} \times \mathbb{T}^3 \to \mathbb{R}^3$$

obeys the estimate

$$|\partial_a \Phi_s^i(t, x)| \leq C_1 c^{C_2 \Xi e_v^{1/2} s} N^{(|a|-L)_+/L} \Xi^{|a|-1}. \qquad (625)$$

In particular, for $|s| \leq \Xi^{-1} e_v^{1/2}$, we have

$$|\partial_a \Phi_s^i(t, x)| \leq C N^{(|a|-L)_+/L} \Xi^{|a|-1}. \qquad (626)$$

Note that these estimates are very similar in character to the estimate of Proposition (17.3) proven for the phase gradients. We will also choose a very similar method of proof, beginning by introducing a dimensionless energy analogous to Definition (17.1).

Definition 18.1. We define the M'th order dimensionless energy of Φ_s to be

$$E_M[\Phi_s] = \sum_{1 \leq |a| \leq M} \sum_{m=1,2,3} \left| \frac{\partial_a \Phi^m}{\Xi^{|a|-1} N^{(|a|-L)_+/L}} \right|^{\frac{2M}{|a|}} \qquad (627)$$

where the summation runs over all multi-indices a with $1 \leq |a| \leq M$.

In terms of the dimensionless energy, the inequality (625) follows from the Gronwall lemma once we have established the following differential inequality.

Lemma 18.1. *There exists a constant $C = C_M$ such that*

$$\left| \frac{d}{ds} E_M[\Phi_s] \right| \leq C \Xi e_v^{1/2} E_M[\Phi_s]. \qquad (628)$$

Proof. Starting from the definition

$$\frac{d\Phi_s}{ds} = v_\epsilon(\Phi_s) \tag{629}$$

of the coarse scale flow, we derive evolution equations for the spatial derivatives of Φ_s^i by differentiating Equation (629). For example, by applying the chain rule, any first derivative $\partial_a \Phi_s^i$, $|a| = 1$ of a component Φ_s^i of Φ_s is coupled to the first derivatives of the other components by the equation

$$\frac{d}{ds} \partial_a \Phi_s^i = \partial_m v_\epsilon^i(\Phi_s) \partial_a \Phi_s^m \tag{630}$$

with an implied summation over the components m. By multiplying this equation by $2\partial_a \Phi_s^i$, summing over i, and applying the bound $\|\nabla v\|_{C^0} \leq \Xi e_v^{1/2}$ we obtain

$$\left| \frac{d}{ds} \sum_{i=1}^3 (\partial_a \Phi_s^i)^2 \right| \leq C \Xi e_v^{1/2} \left(\sum_{i=1}^3 (\partial_a \Phi_s^i)^2 \right) \tag{631}$$

which implies (628) and (625). For higher order derivatives, it becomes convenient to use the dimensionless energy.

To continue, any second order spatial derivative $\partial_a \Phi_s^i$, $a = (a_1, a_2)$ satisfies

$$\frac{d}{ds}(\partial_{a_1} \partial_{a_2} \Phi_s^i) = \partial_{m_1} \partial_{m_2} v_\epsilon^i(\Phi_s) \partial_{a_2} \Phi_s^{m_2} \partial_{a_1} \Phi_s^{m_1} + \partial_{m_2} v_\epsilon^i(\Phi_s) \partial_{a_1} \partial_{a_2} \Phi_s^{m_2} \tag{632}$$

with an implied summation over each index m_1, m_2. From this formula we see that even though $\partial_a \Phi_s$ is initially 0 when $s = 0$ and $\Phi_0(t, x) = (t, x)$, the first of these terms is already of size $\Xi^2 e_v^{1/2}$, which is responsible for the immediate growth of $\partial_a \Phi_s$.

A general higher order derivative $\partial_a \Phi_s$ of order $|a| = |(a_1, \dots, a_A)| = A$ evolves according to the equation

$$\frac{d}{ds}(\partial_a \Phi_s) = \sum_{K=1}^A \sum_{(a^1, \dots, a^K) \in P_K(a)} \partial_{m_1} \dots \partial_{m_K} v_\epsilon \cdot \prod_{i=1}^K \partial_{a^i} \Phi_s^{m_i}. \tag{633}$$

Here $P_K(a)$ denotes the set of ordered K-partitions of a, as explained in Definition (4.1), and each a^i in the product is a multi-index.

The important feature of the above formula is that the number of Φ terms in each product (which is denoted by K) is equal to the number of times that v_ϵ has been differentiated.

Using Equation (633) we can now prove the differential inequality (628).

We first differentiate the individual term in the dimensionless energy

$$\frac{d}{ds}\left|\frac{\partial_a \Phi_s^m}{\Xi^{(|a|-1)}N^{(|a|-L)_+/L}}\right|^{\frac{2M}{|a|}}$$

$$= \left(\frac{\partial_a \Phi_s^m}{\Xi^{(|a|-1)}N^{(|a|-L)_+/L}}\right)^{\left(\frac{2M}{a}-1\right)} \cdot \left(\frac{1}{\Xi^{(|a|-1)}N^{(|a|-L)_+/L}}\frac{d}{ds}(\partial_a \Phi_s^m)\right)$$

which already gives a termwise bound of

$$\left|\frac{d}{ds}\left|\frac{\partial_a \Phi_s^m}{\Xi^{(|a|-1)}N^{(|a|-L)_+/L}}\right|^{\frac{2M}{a}}\right| \leq E_M[\Phi_s]^{1-\frac{|a|}{2M}} \cdot \frac{|\frac{d}{ds}(\partial_a \Phi_s)|}{\Xi^{(|a|-1)}N^{(|a|-L)_+/L}}.$$

$$(634)$$

Then using Equation (633), we can bound this last factor if we can bound each term in the summation

$$\frac{|\frac{d}{ds}(\partial_a \Phi_s)|}{\Xi^{(|a|-1)}N^{(|a|-L)_+/L}}$$

$$\leq \sum_{K-1}^{A}\sum_{(a^1,\ldots,a^K)\in P_K(a)}\frac{|\partial_{m_1}\ldots\partial_{m_K}v_\epsilon \cdot \prod_{i=1}^{K}\partial_{a^i}\Phi_s^{m_i}|}{\Xi^{(|a|-1)}N^{(|a|-L)_+/L}}$$

$$(635)$$

by a constant times $E_M[\Phi_o]^{\frac{|a|}{2M}}$.

As in the proof of the estimate (564) we first apply the counting inequality in Lemma (17.1) with $K-1$, $|a_i|-1$ and $L-1 \geq 0$ to conclude that

$$(K-L)_| + \sum_{i=1}^{K}(|a_i|-L)_+$$

$$= ((K-1)-(L-1))_+ + \sum_{i=1}^{K}((|a_i|-1)-(L-1))_+$$

$$\leq (|a|-L)_+ = \left((K-1)+\sum_{i=1}^{K}(|a_i|-1)-(L-1)\right)_+.$$

$$(636)$$

Using inequality (636), we distribute the powers of N and Ξ for the terms in (635)

$$\frac{|\partial_{m_1}\ldots\partial_{m_K}v_\epsilon \cdot \prod_{i=1}^{K}\partial_{a^i}\Phi_s^{m_i}|}{\Xi^{(|a|-1)}N^{(|a|-L)_+/L}}$$

$$\leq \frac{|\partial_{m_1}\ldots\partial_{m_K}v_\epsilon|}{\Xi^{K-1}N^{(|K|-L)_+/L}}\prod_{i=1}^{k}\frac{|\partial_{a^i}\Phi_s^{m_i}|}{\Xi^{(|a_i|-1)}N^{(|a_i|-L)_+/L}}.$$

Applying the bound (474), the right-hand side can be bounded by

$$\frac{|\partial_{m_1}\ldots\partial_{m_K}v_\epsilon|}{\Xi^{K-1}N^{(|K|-L)_++/L}}\prod_{i=1}^{k}\frac{|\partial_{a^i}\Phi_s^{m_i}|}{\Xi^{(|a_i|-1)}N^{(|a_i|-L)_++/L}}\leq\Xi e_v^{1/2}\cdot\prod_{i=1}^{k}E_M[\Phi_s]^{\frac{|a_i|}{2M}}$$

$$=\Xi e_v^{1/2}E_M[\Phi_s]^{\frac{|a|}{2M}} \qquad (637)$$

which is the estimate we desired to deduce (628) from (635). $\qquad\square$

18.5 Spatial Variations of the Mollified Stress

Here we collect estimates for the mollified stress and its derivatives, starting with the C^0 bound.

Proposition 18.2.

$$\|R_\epsilon\|_{C^0}\leq e_R \qquad (638)$$

which is clear from the definition of R_ϵ as an average value of R.

Using the distortion estimate (626), we can also prove bounds on the derivatives of R_ϵ.

Proposition 18.3.

$$\|\partial_a R_\epsilon\|_{C^0}\leq CN^{(|a|-L)_++/L}\Xi^{|a|}e_R \qquad (639)$$

Proof. The bound

$$\|\partial_a R_{\epsilon_x}\|_{C^0}\leq C_{|a|}N^{(|a|-L)_++/L}\Xi^{|a|}e_R \qquad (640)$$

is clear for

$$R_{\epsilon_x}=\eta_{\epsilon_x}*\eta_{\epsilon_x}*R$$

since $\|\nabla^{|a|}R\|_{C^0}\leq\Xi^{|a|}e_R$ for all $|a|\leq L$ by (243), and from the choice of $\epsilon_x=a\Xi^{-1}N^{-1/L}$ in (613).

Taking a worse constant, we can prove the same estimate for the fully mollified version

$$R_\epsilon(t,x)=\int R_{\epsilon_x}(\Phi_s(t,x))\eta_{\epsilon_t}(s)ds \qquad (641)$$

$$=\int R_{\epsilon_x}(t+s,\Phi_s^i(t,x))\eta_{\epsilon_t}(s)ds \qquad (642)$$

by applying the chain rule and the bounds (626) as follows. We first calculate for any fixed t and s the derivative

$$\partial_a R_{\epsilon_x}(t+s,\Phi_s^i(t,x))=\sum_{K=1}^{A}\sum_{(a^1,\ldots,a^K)\in P_K(a)}\partial_{m_1}\ldots\partial_{m_K}R_{\epsilon_x}\cdot\prod_{i=1}^{K}\partial_{a^i}\Phi_s^{m_i}.$$

$$(643)$$

The terms involving derivatives of Φ_s^i can be estimated using (626) because the time ϵ_t is sufficiently small by (622). By also applying the estimate (640), we can bound each term that appears in the sum by

$$
|\partial_{m_1}\ldots\partial_{m_K} R_{\epsilon_x}\cdot\prod_{i=1}^{K}\partial_{a^i}\Phi_s^{m_i}|
$$

$$
(644)
$$

$$
\leq C(N^{(K-L)_+/L}\Xi^K e_R)\cdot\prod_{i=1}^{K}(N^{(|a_i|-L)_+/L}\Xi^{|a_i|-1})
$$

$$
= CN^{(K-L)_+/L+\sum_i(|a_i|-L)_+/L}\Xi^{K+\sum_{i=1}^{K}(|a_i|-1)}e_R \qquad (645)
$$

$$
= CN^{(K-L)_+/L+\sum_i(|a_i|-L)_+/L}\Xi^{|a|}e_R \qquad (646)
$$

for any ordered K-partition $(a^1,\ldots,a^K)$ of a.

We now recall the estimate

$$
(K-L)_+ + \sum_{i=1}^{K}(|a_i|-L)_+ \leq (|a|-L)_+ \qquad |a_i|, L \geq 1, \qquad (647)
$$

which was proven in (636) using the counting inequality, to see that

$$
|\partial_{m_1}\ldots\partial_{m_K} R_{\epsilon_x}\cdot\prod_{i=1}^{K}\partial_{a^i}\Phi_s^{m_i}| < CN^{(|a|-L)_+/L}\Xi^{|a|}e_R \qquad (648)
$$

for any ordered K partition $(a^1,\ldots,a^K)$ of a. This estimate bounds every term in (643) and therefore completes the proof of (18.3). $\qquad\square$

18.6   Transport Estimates for the Mollified Stress

Here we collect estimates for the material derivative of the mollified stress

$$
\frac{\bar{D}R_\epsilon}{\partial t} = (\partial_t + v_\epsilon^a\partial_a)R_\epsilon^{jl}
$$

as well as its spatial derivatives.

The results of this section are summarized by the following proposition.

Proposition 18.4 (First Material Derivative of the Mollified Stress). *For $k \geq 0$, there exist constants C_k such that*

$$
\|\nabla^k\frac{\bar{D}R_\epsilon}{\partial t}\|_{C^0} \leq C_k N^{(k+1-L)_+/L}\Xi^{k+1}e_v^{1/2}e_R. \qquad (649)
$$

We begin by calculating

$$
(\partial_t + v_\epsilon^a(t,x)\partial_a)R_\epsilon^{jl}(t,x) = \int(\partial_t + v_\epsilon^a(t,x)\partial_a)R_{\epsilon_x}^{jl}(\Phi_s(t,x))\eta_{\epsilon_t}(s)ds \quad (650)
$$

$$
= \int DR_{\epsilon_x}^{jl}(\Phi_s(t,x))D\Phi_s(t,x)(\partial_t + v_\epsilon^a(t,x)\partial_a)\eta_{\epsilon_t}(s)ds
$$

$$
(651)
$$

using the chain rule.

Here and in the remainder of the section, $DF(t, x)$ denotes the derivative of F in the (t, x) variables. Thus

$$DFX^\mu \partial_\mu = DF(t, x)(X^\mu(t, x)\partial_\mu) \tag{652}$$
$$= X^\mu(t, x)\partial_\mu F(t, x) \tag{653}$$

for any 4-vector $X^\mu(t, x)\partial_\mu$.

Because Φ_s has been defined to be the flow map of the vector field $\frac{\bar{D}}{\partial t} = (\partial_t + v_\epsilon^a(t, x)\partial_a)$, we can show that applying the flow Φ_s to $\frac{\bar{D}}{\partial t}$

$$D\Phi_s(t, x)[\partial_t + v_\epsilon^a(t, x)\partial_a] = \partial_t + v_\epsilon^a(\Phi_s(t, x))\partial_a \tag{654}$$

simply moves the vector field $(\partial_t + v_\epsilon^a(t, x)\partial_a)$ forward along the flow.

As a consequence, we can show the following lemma.

Lemma 18.2 (Derivatives and Averages Along the Flow Commute). *For all C^1 functions F on $\mathbb{R} \times \mathbb{T}^3$ we have an equality*

$$(\partial_t + v_\epsilon^a(t, x)\partial_a)[F(\Phi_s(t, x))] = \frac{\bar{D}F}{\partial t}(\Phi_s(t, x)). \tag{655}$$

In particular,

$$(\partial_t + v_\epsilon^a(t, x)\partial_a)R_\epsilon^{jl}(t, x) = \int \frac{\bar{D}R_{\epsilon_x}}{\partial t}(\Phi_s(t, x))\eta_{\epsilon_t}(s)ds \tag{656}$$

where

$$\int \frac{\bar{D}R_{\epsilon_x}}{\partial t}(\Phi_s(t, x))\eta_{\epsilon_t}(s)ds$$
$$= \int [\partial_t R_{\epsilon_x}^{jl}(\Phi_s(t, x)) + v_\epsilon^a(\Phi_s(t, x))[\partial_a R_{\epsilon_x}^{jl}](\Phi_s(t, x))]\eta_{\epsilon_t}(s)ds \tag{657}$$
$$= \int DR_{\epsilon_x}(\Phi_s(t, x))[(\partial_t + v_\epsilon^a(\Phi_s(t, x))\partial_a)]\eta_{\epsilon_t}(s)ds. \tag{658}$$

We give a proof of this (crucial) calculation in the following section. We then apply the lemma in Section (18.6.2) to establish the estimates in Proposition (18.4).

18.6.1 Derivatives and Averages along the Flow Commute

The purpose of this section is to establish Lemma (18.2), which is a consequence of the following more general fact, applied to the 4-vector fields

$$X^\mu(t, x)\partial_\mu = \partial_t + v_\epsilon^a(t, x)\partial_a$$
$$Y^\mu(t, x)\partial_\mu = \partial_t + v_\epsilon^a(t, x)\partial_a$$

which, in our case, are equal to each other.

Lemma 18.3. *Let* $Y^\mu(t,x)\partial_\mu : \mathbb{R} \times \mathbb{T}^3 \to \mathbb{R} \times \mathbb{R}^3$ *be any smooth vector field on* $\mathbb{R} \times \mathbb{T}^3$, *let* $\Phi_s(t,x)$ *denote the time s flow of Y. That is,* $\Phi_s(t,x)$ *solves the ODE*

$$\frac{d\Phi_s^\mu}{ds} = Y^\mu(\Phi_s(t,x)) \tag{659}$$

$$\Phi_0(t,x) = (t,x). \tag{660}$$

Suppose that X commutes with Y in the sense that the commutator

$$[X,Y]^\nu \partial_\nu = (X^\mu \partial_\mu Y^\nu - Y^\mu \partial_\mu X^\nu)\partial_\nu \tag{661}$$

vanishes.

Then X also commutes with pullback along the flow of Y,

$$D\Phi_s(t,x)(X^\mu(t,x)\partial_\mu) = X^\mu(\Phi_s(t,x))\partial_\mu.$$

That is, for any smooth function F on $\mathbb{R} \times \mathbb{T}^3$, *we have*

$$X^\mu(\Phi_s(t,x))(\partial_\mu F)(\Phi_s(t,x)) = X^\mu(t,x)\partial_\mu [F(\Phi_s(t,x))]. \tag{662}$$

Proof. In order to compare

$$X^\mu(\Phi_s(t,x))(\partial_\mu F)(\Phi_s(t,x)) = DF(\Phi_s(t,x))[X^\mu(\Phi_s(t,x))\partial_\mu] \tag{663}$$

$$= DF(\Phi_s(t,x))D\Phi_s(t,x)[D\Phi_s(t,x)]^{-1}X^\mu(\Phi_s(t,x))\partial_\mu \tag{664}$$

and

$$X^\mu(t,x)\partial_\mu[F(\Phi_s(t,x))] - DF(\Phi_s(t,x))D\Phi_s(t,x)[X^\mu(t,x)\partial_\mu] \tag{665}$$

$$= DF(\Phi_s(t,x))D\Phi_s(t,x)[D\Phi_0(t,x)]^{-1}X^\mu(\Phi_0(t,x))\partial_\mu \tag{666}$$

it suffices to show that the vector field

$$Z_s^\mu(t,x)\partial_\mu = [D\Phi_s(t,x)]^{-1}X^\mu(\Phi_s(t,x))\partial_\mu, \tag{667}$$

which is characterized by the equation

$$\partial_\mu \Phi_s^\nu(t,x)Z_s^\mu(t,x)\partial_\nu = X^\nu(\Phi_s(t,x))\partial_\nu, \tag{668}$$

satisfies

$$\dot{Z}_s^\mu \partial_\mu = \frac{\partial}{\partial s}Z_s^\mu(t,x)\partial_\mu = 0. \tag{669}$$

To show (669), we differentiate (668) and apply (659) to see that

$$\partial_\mu \Phi_s^\nu(t,x)\dot{Z}_s^\mu(t,x)\partial_\nu = -\frac{\partial}{\partial s}[\partial_\mu \Phi_s^\nu(t,x)]Z_s^\mu(t,x) + \frac{\partial}{\partial s}[X^\nu(\Phi_s(t,x))]\partial_\nu \tag{670}$$

$$= -\frac{\partial}{\partial s}[\partial_\mu \Phi_s^\nu(t,x)]Z_s^\mu(t,x) + Y^\mu(\Phi_s(t,x))[\partial_\mu X^\nu](\Phi_s(t,x))\partial_\nu \tag{671}$$

$$\frac{\partial}{\partial s}[\partial_\mu \Phi_s^\nu(t,x)] = \partial_\alpha Y^\nu(\Phi_s(t,x))\partial_\mu \Phi_s^\alpha(t,x). \tag{672}$$

Then comparing (672) and (668) we obtain

$$
\begin{aligned}
\partial_\mu \Phi_s^\nu(t,x) \dot{Z}_s^\mu(t,x) \partial_\nu &= -X^\alpha(\Phi_s)[\partial_\alpha Y^\nu](\Phi_s(t,x))\partial_\nu \\
&\quad + Y^\mu(\Phi_s(t,x))[\partial_\mu X^\nu](\Phi_s(t,x))\partial_\nu
\end{aligned}
\tag{673}
$$

$$
= -[X,Y]^\nu(\Phi_s(t,x))\partial_\nu = 0.
\tag{674}
$$

$\square$

The reader may recognize that the above argument is essentially an ODE proof that $e^{sA}B = Be^{sA}$ for any pair of commuting matrices A, B and any $s \in \mathbb{R}$, except that e^{sA} has been replaced by the pullback operator $\circ\Phi_s$, and B has been replaced by the differential operator $X^\mu\partial_\mu$.

18.6.2 Material Derivative Bounds for the Mollified Stress

In this section, we prove Proposition (18.4) by estimating the spatial derivatives of

$$
\frac{\bar{D}R_\epsilon}{\partial t} = \int \frac{\bar{D}R_{\epsilon_x}}{\partial t}(\Phi_s(t,x))\eta_{\epsilon_t}(s)ds
\tag{675}
$$

as well as its spatial derivatives.

Proposition 18.5 (Material Derivative Bounds for the Spatially Mollified Stress). *There exist constants depending on k such that*

$$
\|\nabla^k \frac{\bar{D}R_\epsilon}{\partial t}\|_{C^0} \leq C_k N^{(k+1-L)_+/L} \Xi^{k+1} e_v^{1/2} e_R.
\tag{676}
$$

As a first step, we estimate the tensor field

$$
\frac{\bar{D}R_{\epsilon_x}^{jl}}{\partial t} = (\partial_t + v_\epsilon^a(t,x)\partial_a)R_{\epsilon_x}^{jl}(t,x)
\tag{677}
$$

and its spatial derivatives, which are summarized by the following lemma.

Proposition 18.6 (Bounds for $\frac{\bar{D}R_{\epsilon_x}}{\partial t}$). *For every $k \geq 0$ there is a constant C_k such that*

$$
\|\nabla^k \frac{\bar{D}R_{\epsilon_x}}{\partial t}\|_{C^0} \leq C_k N^{(k+1-L)_+/L} \Xi^{k+1} e_v^{1/2} e_R.
\tag{678}
$$

Given Proposition (18.6), the estimate in Proposition (18.5) follows just as in Section (18.5). To summarize the argument of that section, each spatial derivative up to order $L - 1$ costs at most Ξ, even if it is taken on the flow Φ_s, and beyond $L-1$ derivatives, the cost of each spatial derivative increases to $\Xi N^{1/L}$. No exponential factors appear in the estimates for derivatives of Φ_s because we have checked that $s \leq \epsilon_t = N^{-1}\Xi^{-1}e_R^{-1/2} < \Xi^{-1}e_v^{-1/2}$.

126

Estimating $\frac{\bar{D}R_{\epsilon_x}}{\partial t}$ requires us to commute the operator

$$\frac{\bar{D}}{\partial t} = \partial_t + v_\epsilon^a(t, x)\partial_a$$

with the spatial mollifier

$$\eta_{\epsilon+\epsilon}* = \eta_{\epsilon_x} * \eta_{\epsilon_x} * .$$

This commutation leads to an expansion of $\frac{\bar{D}R_{\epsilon_x}}{\partial t}$ into three terms.[17]

$$\frac{\bar{D}R_{\epsilon_x}}{\partial t} = \eta_{\epsilon+\epsilon} * \frac{\bar{D}R}{\partial t} + [v_\epsilon^a(t, x)\partial_a, \eta_{\epsilon_x}*](\eta_{\epsilon_x} * R) \tag{679}$$
$$+ \eta_{\epsilon_x} * ([v_\epsilon^a(t, x)\partial_a, \eta_{\epsilon_x}*]R)$$
$$= A_{(I)}(t, x) + A_{(II)}(t, x) + A_{(III)}(t, x) \tag{680}$$

We first express the commutator terms as

$$A_{(II)}(t, x) = \int [v_\epsilon^a(t, x) - v_\epsilon^a(t, x+h)]\partial_a(\eta_{\epsilon_x} * R)(t, x)\eta_{\epsilon_x}(h)dh \tag{681}$$

$$= \int_0^1 \int \partial_i v_\epsilon^a(t, x+sh)\partial_a(\eta_{\epsilon_x} * R)(t, x)h^i\eta_{\epsilon_x}(h)dh \tag{682}$$

$$= \epsilon_x \int_0^1 \int \partial_i v_\epsilon^a(t, x+sh)\partial_a(\eta_{\epsilon_x} * R)(t, x)\tilde{\eta}_{\epsilon_x}^i(h)dh \tag{683}$$

with $\epsilon_x = \Xi^{-1}N^{-1/L}$. From this expression, we can conclude that

$$\|\nabla^k A_{(II)}\|_{C^0}$$
$$\leq C_k\Xi^{-1}N^{-1/L} \sum_{|a_1|+|a_2|=k} \|\nabla^{|a_1|+1}v_\epsilon\|_{C^0}\|\nabla^{|a_2|+1}(\eta_{\epsilon_x} * R)\|_{C^0} \tag{684}$$

$$\leq C\Xi^{-1}N^{-1/L}.$$
$$\sum_{|a_1|+|a_2|=k} \left(N^{(|a_1|+1-L)_+/L}\Xi^{|a_1|+1}e_v^{1/2}\right)\left(N^{(|a_2|+1-L)_+/L}\Xi^{|a_2|+1}e_R\right)$$
$$\tag{685}$$

$$\leq C\Xi^{k+1}e_v^{1/2}e_R \sum_{|a_1|+|a_2|=k} (N^{[(|a_1|-(L-1))_+ +(|a_2|-(L-1))_+ -1]/L}) \tag{686}$$

$$\leq CN^{(k+1-L)_+/L}\Xi^{k+1}e_v^{1/2}e_R. \tag{687}$$

where in the last line we have used the counting inequality in Lemma (17.1). We can similarly treat

$$A_{(III)}(t, x) = \eta_{\epsilon_x} * ([v_\epsilon^a(t, x)\partial_a, \eta_{\epsilon_x}*]R) \tag{688}$$

$$[v_\epsilon^a(t, x)\partial_a, \eta_{\epsilon_x}*]R = \int_0^1 \int \partial_i v_\epsilon^a(t, x+sh)\partial_a R(t, x)h^i\eta_{\epsilon_x}(h)dh. \tag{689}$$

---

[17]Note that these indices are Roman numerals, and should not be confused with the indices I for the individual waves V_I.

The commutator can be differentiated $L - 1$ times to obtain

$$\|\nabla^k [v_\epsilon^a(t,x)\partial_a, \eta_{\epsilon_x} *]R\|_{C^0} \le C_k \epsilon_x \Xi^{2+k} e_v^{1/2} e_R \tag{690}$$

$$\le C_k N^{-1/L} \Xi^{1+k} e_v^{1/2} e_R. \tag{691}$$

Taking more than $L - 1$ derivatives of $A_{(III)}$ incurs a loss of $\epsilon_x^{-1} \lesssim \Xi N^{1/L}$, giving us a bound of

$$\|\nabla^k A_{(III)}\|_{C^0} \le N^{(k-L)_+/L} \Xi^{k+1} e_v^{1/2} e_R \tag{692}$$

$$\le N^{(k+1-L)_+/L} \Xi^{k+1} e_v^{1/2} e_R. \tag{693}$$

In order to estimate the term

$$\eta_{\epsilon+\epsilon} * \frac{\bar{D}R}{\partial t} = \eta_{\epsilon+\epsilon} * [(\partial_t + v_\epsilon \cdot \nabla)R]$$

we use the assumed bounds on $(\partial_t + v \cdot \nabla)R$ by writing

$$\eta_{\epsilon+\epsilon} * \frac{\bar{D}R}{\partial t} = \eta_{\epsilon+\epsilon} * [(\partial_t + v \cdot \nabla)R] + \eta_{\epsilon+\epsilon} * [(v_\epsilon - v) \cdot \nabla R] \tag{694}$$

$$= A_{(I,I)} + A_{(I,II)}. \tag{695}$$

Then

$$\|\nabla^k A_{(I,I)}\|_{C^0} \le C_k N^{(k+1-L)_+/L} \Xi^{k+1} e_v^{1/2} e_R \tag{696}$$

since the first $L - 1$ derivatives do not need to fall on the mollifier, and beyond $L - 1$, each derivative taken costs $\epsilon_x^{-1} \lesssim \Xi N^{1/L}$.

For $A_{(I,II)}$ we have

$$(v_\epsilon - v) \cdot \nabla R = \int [v^a(t, x + h) - v^a(t,x)]\partial_a R(t,x)\eta_{\epsilon+\epsilon}(h)dh \tag{697}$$

$$= \int_0^1 \int \partial_i v^a(t, x + sh)\partial_a R(t,x)h^i \eta_{\epsilon+\epsilon}(h)dh \tag{698}$$

which can be differentiated up to $L - 1$ times at a cost of Ξ per derivative

$$\|\nabla^k[(v_\epsilon - v) \cdot \nabla R]\|_{C^0} \le C_k N^{-1/L} \Xi^{k+1} e_v^{1/2} e_R \tag{699}$$

so that each derivative taken beyond $L - 1$ costs $\epsilon_x^{-1} \lesssim \Xi N^{1/L}$ as it falls on the mollifier, giving

$$\|\nabla^k A_{(I,II)}\|_{C^0} \le C_k N^{(k-L)_+/L} \Xi^{k+1} e_v^{1/2} e_R \tag{700}$$

$$\le N^{(k+1-L)_+/L} \Xi^{k+1} e_v^{1/2} e_R. \tag{701}$$

This finishes the proof of Proposition (18.6).

18.6.3 Second Time Derivative of the Mollified Stress along the Coarse Scale Flow

In this section, we establish bounds for the second material derivative

$$\frac{\bar{D}^2 R_\epsilon}{\partial t^2} = (\partial_t + v_\epsilon \cdot \nabla)^2 R_\epsilon \tag{702}$$

of the mollified stress R_ϵ, along with its spatial derivatives. These estimates will be necessary to control the material derivative $\frac{\bar{D}Q_T^{jl}}{\partial t}$ where Q_T is the part of the new stress that arises from the Transport term

$$\partial_j Q_T^{jl} = (\partial_t + v_\epsilon^j \partial_j) V^l \tag{703}$$

$$= \sum_I e^{i\lambda \xi_I} [(\partial_t + v_\epsilon^j \partial_j) \tilde{v}_I^l]. \tag{704}$$

If we recall that R_c was constructed by mollifying a time ϵ_t along the coarse scale flow

$$R_c(t, x) = \int R_{\epsilon_x}(\Phi_s(t, x)) \eta_{\epsilon_t}(s) ds \tag{705}$$

then by analogy with standard mollification we expect the estimates for the second material derivative $\frac{D^2}{\partial t^2} R_\epsilon$ to be a factor ϵ_t^{-1} worse than the bounds on the first material derivative $\frac{DR_\epsilon}{\partial t}$.

To see that this expectation is correct, we apply Lemma (18.2) to see that

$$(\partial_t + v_\epsilon^a(t, x) \partial_a) R_\epsilon^{jl}(t, x) = \int \frac{\bar{D} R_{\epsilon_x}}{\partial t} (\Phi_s(t, x)) \eta_{c_t}(s) ds \tag{706}$$

$$= \int \frac{d}{ds} [R_{\epsilon_x}(\Phi_s(t, x))] \eta_{\epsilon_t}(s) ds \tag{707}$$

$$= -\int R_{\epsilon_x}(\Phi_s(t, x)) \eta'_{\epsilon_t}(s) ds. \tag{708}$$

The proof of Lemma (18.2) has nothing to do with the choice of the function η_{ϵ_t}, and we can express the second material derivative of R_ϵ as

$$\frac{\bar{D}^2 R_\epsilon}{\partial t^2} = -\int \frac{\bar{D} R_{\epsilon_x}}{\partial t} (\Phi_s(t, x)) \eta'_{\epsilon_t}(s) ds. \tag{709}$$

which is essentially the kind of average estimated in Section (18.6.2), except that the bound on $\|\eta'_{\epsilon_t}(s)\|_{L_s^1}$ is worse by a factor

$$\epsilon_t^{-1} = N \Xi e_R^{1/2}.$$

The same methods from Sections (18.6.2) and (18.5) then imply the following bounds.

129

Proposition 18.7 (Second Material Derivative of the Mollified Stress).
For $k \geq 0$, there exist constants C_k such that

$$\|\nabla^k \frac{\bar{D}^2 R_\epsilon}{\partial t^2}\|_{C^0} \leq C_k N^{1+(k+1-L)_+/L} \Xi^{k+2} e_v^{1/2} e_R^{3/2}. \qquad (710)$$

Now that we have established the bounds of Proposition (18.7), we should check that this estimate is acceptable for the purpose of proving the Main Lemma (10.1).

18.6.4 An Acceptability Check

The purpose of mollifying R^{jl} along the coarse scale flow is to ensure that the material derivative of the Transport term Q_T^{jl} obeys good bounds. Let us now calculate the bounds we expect for $\frac{\bar{D} Q_T^{jl}}{\partial t}$ and check how they compare to the requirements of Conjecture (10.1).

Recall that the first term in the parametrix expansion for the solution to

$$\partial_j Q_{T,I}^{jl} = e^{i\lambda\xi_I} u_{T,I,(1)}^l \qquad (711)$$

$$u_{T,I,(1)}^l = (\partial_t + v_\epsilon^j \partial_j)\tilde{v}_I^l \qquad (712)$$

is of the form

$$Q_{T,I,(1)}^{jl} = \frac{1}{\lambda} e^{i\lambda\xi_I} q_{T,I,(1)}^{jl} \qquad (713)$$

$$q_{T,I,(1)}^{jl} = i^{-1} q^{jl}(\nabla\xi_I)[u_{T,I,(1)}] \qquad (714)$$

where $q^{jl}(\nabla\xi_I)[u]$ is the solution to $\partial_j \xi_I q^{jl} = u^l$ described in Section (6), which depends linearly on u.

To leading order in λ,

$$u_{T,I,(1)}^l \approx \frac{\bar{D}}{\partial t} v_I^l \qquad (715)$$

$$v_I^l = a_I^l + i b_I^l \qquad (716)$$

$$b_I^l = \eta_{k_4}(t)\psi_k(t,x)e^{1/2}(t)\gamma_I(\nabla\xi_k,\varepsilon)P_I^\perp(\nabla\xi_{\sigma I})^l \qquad (717)$$

$$\varepsilon^{jl} = \frac{-\mathring{R}_\epsilon^{jl}}{e(t)} \qquad (718)$$

$$a_I = -\frac{(\nabla\xi_I)}{|\nabla\xi_I|} \times b_I \qquad (719)$$

which means that one of the terms appearing in $u_{T,I,(1)}^l$ involves $\frac{\bar{D}}{\partial t} R_\epsilon$ when the material derivative hits the ε inside of γ_I. This term has size

$\Xi e_v^{1/2} e_R^{1/2}$ because it costs $\Xi e_v^{1/2}$ to take the first material derivative of R_ϵ, and therefore gives a contribution to the Transport term of size

$$|Q_{T,I,(1)}^{jl}| = \frac{e_v^{1/2} e_R^{1/2}}{N} + |\text{ other terms }|. \tag{720}$$

It will ultimately turn out that some of the other terms will be larger, but for now we will focus on the term involving $\frac{\bar{D}R_\epsilon}{\partial t}$.

To establish Lemma (10.1), we will have to verify a bound on

$$(\partial_t + v_1 \cdot \nabla) Q_{T,I,(1)}^{jl} = (\partial_t + v \cdot \nabla) Q_{T,I,(1)}^{jl} + V \cdot \nabla Q_{T,I,(1)}^{jl}$$

and its first $L - 1$ spatial derivatives.

During this verification, we approximate $(\partial_t + v \cdot \nabla) = \frac{\bar{D}}{\partial t} + (v - v_\epsilon) \cdot \nabla$ because $\frac{\bar{D}}{\partial t}$ will not hit the phase functions. Then, in order to bound $\frac{\bar{D} Q_{T,I,(1)}^{jl}}{\partial t}$ we will take a second material derivative of R_ϵ. This second derivative will cost a factor ϵ_t^{-1}, leading to the bounds

$$|\frac{\bar{D}}{\partial t} Q_{T,I,(1)}^{jl}| \leq \epsilon_t^{-1} \frac{e_v^{1/2} c_R^{1/2}}{N} + |\text{ other terms }| \tag{721}$$

$$\leq C \Xi e_v^{1/2} e_R + |\text{ other terms }|. \tag{722}$$

In the ideal case where $(\Xi', c_n', e_R') = (CN\Xi, c_R, \frac{e_v^{1/2} e_R^{1/2}}{N})$, the right-hand side is exactly the benchmark necessary to be deemed acceptable. Namely,

$$C\Xi e_v^{1/2} e_R = \Xi'(e_v')^{1/2} e_R' \qquad \text{for Conjecture (10.1).}$$

To prove Lemma (10.1), we actually only need to verify that

$$|\frac{\bar{D}}{\partial t} Q_{T,I,(1)}^{jl}| \trianglelefteq CN\Xi e_R^{1/2} \left(\frac{e_v^{1/2}}{e_R^{1/2} N} \right)^{1/2} e_R \tag{723}$$

$$= C \left(\frac{e_v^{1/2}}{e_R^{1/2} N} \right)^{-1/2} (\Xi e_v^{1/2} e_R) \tag{724}$$

which is an even more forgiving benchmark.

For the purpose of proving the Main Lemma, Lemma (10.1), we can therefore afford at other points in the argument to be slightly wasteful in the estimates.

19 Accounting for the Parameters and the Problem with the High-High Term

Although many estimates have been proven so far in the argument, only a few of the parameters have been chosen.

Here we give a list of all the parameters, and summarize their current status:

- The parameter

$$\epsilon_x = a_R \Xi^{-1} N^{-1/L} \tag{725}$$

used to mollify R in space has been chosen in Line (613) of Section (18.3) with $a_R \leq 1$ a constant depending only on the parameter L in the Main Lemma, Lemma (10.1).

- The parameter

$$\epsilon_t = c \Xi^{-1} N^{-1} e_R^{-1/2} \tag{726}$$

used to mollify R_{ϵ_x} along the flow has been chosen in Line (621) of Section (18.3) with $c \leq 1$ an absolute constant.

- The parameter ϵ_v used to mollify the velocity field v_ϵ in space has the form

$$\epsilon_v = a_v \Xi^{-1} N^{-1/L} \tag{727}$$

where the constant $a_v \leq 1$ was chosen in Section (15) Line (471) and depends only on the parameter L in the Main Lemma, Lemma (10.1).

- The lifespan parameter τ is of the form

$$\tau = b \Xi^{-1} e_v^{-1/2} \tag{728}$$

for a dimensionless parameter b that has not yet been chosen.

Although b has not been chosen, b is required to satisfy an upper bound of the form

$$b \trianglelefteq b_0. \tag{729}$$

The number $b_0 < 1$ above is an absolute constant, already specified, that will ensure the conditions specified in Proposition (17.1) are satisfied. The parameter τ does not depend in any way on the choice of the mollifying parameter ϵ_v for the velocity, which will be apparent shortly as b is chosen in Line (732).

- The parameter λ used to make sure the phase functions are high frequency is of the form

$$\lambda = B_\lambda \Xi N \tag{730}$$

where $B_\lambda \geq 1$ is the largest constant in the book, which will be chosen at the very end of the proof. The choice of B_λ depends only on the parameters $L \geq 2$ and $\eta > 0$ in the statement of the Main Lemma, Lemma (10.1).

For the sake of the logical sequence of the argument, we will take this opportunity to choose the parameter τ, and then provide a motivation for the choice. Namely, we set

$$b = b_0 \left(\frac{e_v^{1/2}}{B_\lambda e_R^{1/2} N} \right)^{1/2} \tag{731}$$

$$= b_0 B_\lambda^{-1/2} \left(\frac{e_v^{1/2}}{e_R^{1/2} N} \right)^{1/2} \tag{732}$$

so that

$$\tau = b_0 B_\lambda^{-1/2} \left(\frac{e_v^{1/2}}{e_R^{1/2} N} \right)^{1/2} \Xi^{-1} e_v^{-1/2}. \tag{733}$$

Note that $N \geq \left(\frac{e_v}{e_R} \right)^{1/2}$, so that $b \leq b_0$ as required in (729).

The motivation for the above choice is the following. Consider the High-High term and the Transport term, which solve the equations

$$\partial_j Q_H^{jl} = \lambda \sum_{J \neq I} e^{i\lambda(\xi_I + \xi_J)} [v_I \times (|\nabla \xi_J| - 1)v_J + v_J \times (|\nabla \xi_I| - 1)v_I]$$
$$+ \text{ lower order terms} \tag{734}$$

$$\partial_j Q_T^{jl} = e^{i\lambda \xi_I} (\partial_t + v_\epsilon^j(t, x)\partial_j)v_I^l + \text{ lower order terms} \tag{735}$$

and recall that v_I is given by $a_I + ib_I$, where

$$b_I^l = \eta \left(\frac{t - t(I)}{\tau} \right) \psi_k(t, x) e^{1/2}(t) \gamma_I (\nabla \xi_k, \frac{R_\epsilon}{c}) P_I^\perp (\nabla \xi_{\sigma I})^l \tag{736}$$

and $a_I = -\frac{(\nabla \xi_I) \times}{|\nabla \xi_I|} b_I$ is obtained by a $\pi/2$ rotation of b_I in $\langle \nabla \xi_I \rangle^\perp$.

From the bounds we have established, it is already clear that

$$\|b_I\|_{C^0} \leq C e_R^{1/2} \tag{737}$$

uniformly in I, where the smallness comes from the function $e^{1/2}(t)$, and all the other factors in (736) are bounded.

The solution Q_H to (734) will not be any smaller than the first term in the parametrix expansion for any of its terms, which gains a factor of $1/\lambda$ compared to the right-hand side. At each time, there are only a bounded number (fewer than $(2 \times 2^3 \times 12)^2$) interaction terms which are nonzero. Therefore, we expect a bound

$$|Q_H| \leq C e_R \max_I \sup_{|t-t(I)| \leq \tau, x \in \mathbb{T}^3} (|\nabla \xi_I|(t, x) - 1) \tag{738}$$

which is only smaller than e_R if the phase gradients are very close to their initial size 1 in absolute value.

According to Proposition (17.1), we have a bound

$$||\nabla \xi_I| - 1| \leq Ab \tag{739}$$

(which we remark is dimensionless), giving

$$|Q_H| \leq Cbe_R. \tag{740}$$

Therefore b must be chosen small so that the estimate for $\|Q_H\|_{C^0}$ will guarantee that Q_H is smaller than the stress from the previous stage.

Unfortunately, choosing a short lifespan forces the Transport term to be large, since the material derivative hits the time cutoff function $\eta(\frac{t-t(I)}{\tau})$ in (736). Therefore we expect an estimate

$$\|(\partial_t + v_\epsilon^j \partial_j)v_I\|_{C^0} \leq C\tau^{-1}e_R^{1/2} \tag{741}$$

which leads to an expected estimate for Q_T of

$$|Q_T| \leq C\frac{\tau^{-1}e_R^{1/2}}{B_\lambda \Xi N} \tag{742}$$

$$\leq Cb^{-1}\frac{e_v^{1/2}e_R^{1/2}}{B_\lambda N}. \tag{743}$$

If we had been able to choose b to be a constant, then the transport term Q_T could be made to have size $\frac{e_v^{1/2}e_R^{1/2}}{N}$ which is what is required for regularity up to $1/3$. Unfortunately, the best we can do is optimize b in order to balance the terms (740) and (743), which leads to the choice of b in (732).

Now the only parameter that remains to be chosen is B_λ, which will be chosen at the very end of the argument.

Part VI

Construction of Regular Weak Solutions: Estimating the Correction

In order to estimate the corrections P and V and their spatial and material derivatives for the proof of Lemma (10.1), and to prepare to estimate each term that composes the new stress, we begin by proving estimates for the vector amplitudes v_I.

Let us recall again that v_I is given by $a_I + ib_I$, where

$$b_I^l = \eta\left(\frac{t - t(I)}{\tau}\right)\psi_k(t, x)e^{1/2}(t)\gamma_I(\nabla\xi_k, \varepsilon)P_I^\perp(\nabla\xi_{\sigma I})^l \tag{744}$$

$$\varepsilon^{jl} = \frac{-\mathring{R}_\epsilon^{jl}}{e(t)} \tag{745}$$

and the real part

$$a_I = -\frac{(\nabla\xi_I)}{|\nabla\xi_I|} \times b_I \tag{746}$$

is obtained by a $\pi/2$ rotation of b_I in the plane $\langle\nabla\xi_I\rangle^\perp$.

We need to estimate spatial derivatives of v_I, and spatial derivatives of its first two material derivatives. To begin this process, we start with the coefficients γ_I.

20 Bounds for Coefficients from the Stress Equation

Recall that the coefficients γ_I are defined implicitly by the equation

$$\sum_{I\in\vec{k}\times\mathbb{F}} A(\nabla\xi_k)_J^I\gamma_I^2 = \left(\frac{\delta^{jl}}{2n} + \varepsilon(R_\epsilon)^{jl}\right)\partial_j\xi_J\partial_l\xi_J \tag{747}$$

of Section (7.3.6). The fact that this equation can be solved for γ_I was ensured in the discussion of Section (18.3). In order to estimate the derivatives of γ_I, we can differentiate Equation (747).

As a preliminary measure, we estimate the derivatives for

$$\varepsilon^{jl} = \frac{-\mathring{R}_\epsilon^{jl}}{e(t)}. \tag{748}$$

These are summarized by the following proposition.

Proposition 20.1 (Bounds for ε).

$$\|\nabla^k \varepsilon\|_{C^0} \leq C_k N^{(k-L)_+/L} \Xi^k \tag{749}$$

$$\|\nabla^k \frac{\bar{D}\varepsilon}{\partial t}\|_{C^0} \leq C_k N^{(k+1-L)_+/L} \Xi^{k+1} e_v^{1/2} \tag{750}$$

$$\|\nabla^k \frac{\bar{D}^2 \varepsilon}{\partial t^2}\|_{C^0} \leq C_k N^{1+(k+1-L)_+/L} \Xi^{k+2} e_v^{1/2} e_R^{1/2} \tag{751}$$

Note that these only differ from the bounds (18.3), (18.4) and (18.7) by a factor of e_R^{-1}.

Proof. The proof of Proposition (20.1) proceeds by studying the equations

$$e(t)\varepsilon^{jl} = -\mathring{R}_\epsilon^{jl} \tag{752}$$

$$e(t)\frac{\bar{D}\varepsilon^{jl}}{\partial t} = -\frac{\bar{D}\mathring{R}_\epsilon^{jl}}{\partial t} - e'(t)\varepsilon^{jl} \tag{753}$$

$$e(t)\frac{\bar{D}^2 \varepsilon^{jl}}{\partial t^2} = -\frac{\bar{D}^2 \mathring{R}_\epsilon^{jl}}{\partial t^2} - 2e'(t)\frac{\bar{D}\varepsilon^{jl}}{\partial t} - e''(t)\varepsilon^{jl}. \tag{754}$$

The bound (749) follows by differentiating (752), and applying the bounds of Proposition (18.3) and the lower bound

$$e(t) \geq K e_R.$$

The bound (750) follows by differentiating (753) in space, and applying the bounds (749), (250) and Proposition (18.4). The point is that the first transport derivative always costs a factor $\Xi e_v^{1/2}$ in the estimates, and each spatial derivative costs a factor of Ξ until the total order of differentiation exceeds L, at which point one obtains a larger cost of $N^{1/L}\Xi$ per derivative.

The proof of (751) proceeds similarly by differentiating (754) and applying the bounds from (250), (749), (750), and Proposition (18.7). The equation for the second material derivative has the form

$$\nabla^k \frac{\bar{D}^2 \varepsilon^{jl}}{\partial t^2} = e^{-1}(t)\left\{-\nabla^k \frac{\bar{D}^2 \mathring{R}_\epsilon^{jl}}{\partial t^2} - 2e'(t)\nabla^k \frac{\bar{D}\varepsilon^{jl}}{\partial t} - e''(t)\nabla^k \varepsilon^{jl}\right\}$$

$$\tag{755}$$

$$\|\nabla^k \frac{\bar{D}^2 \varepsilon^{jl}}{\partial t^2}\|_{C^0} \leq [N\Xi e_R^{1/2}] \cdot [N^{(k+1-L)_+/L}\Xi^{k+1}e_v^{1/2}]$$
$$+ [\Xi e_v^{1/2}][N^{(k+1-L)_+/L}\Xi^{k+1}e_v^{1/2}] \tag{756}$$
$$+ [\Xi e_v^{1/2}]^2[N^{(k-L)_+}\Xi^k]$$

$$= A_{(I)} + A_{(II)} + A_{(III)}. \tag{757}$$

Counting powers of N, we see that $A_{(III)} < A_{(II)}$, and the fact that $A_{(II)} < A_{(I)}$ follows from $N \geq \left(\frac{e_v}{e_R}\right)^{1/2}$. $\qquad\square$

We can now differentiate Equation (747) in order to prove the following.

Proposition 20.2 (Bounds for γ_I). *The coefficients $\gamma_I = \gamma_I(\nabla\xi_k, \varepsilon)$ satisfy the bounds*

$$\|\nabla^k \gamma_I\|_{C^0} \leq C_k N^{(k+1-L)_+/L} \Xi^k \tag{758}$$

$$\|\nabla^k \frac{\bar{D}\gamma_I}{\partial t}\|_{C^0} \leq C_k N^{(k+1-L)_+/L} \Xi^{k+1} e_v^{1/2} \tag{759}$$

$$\|\nabla^k \frac{\bar{D}^2 \gamma_I}{\partial t^2}\|_{C^0} \leq C_k N^{1+(k+1-L)_+/L} \Xi^{k+2} e_v^{1/2} e_R^{1/2}. \tag{760}$$

Since the proof is a routine application of the chain rule, we will only give a schematic outline of which terms appear in order to avoid clutter.

Proof. The proof proceeds either by differentiating (747), or equivalently applying the chain rule to the implicit function

$$\gamma_I(p, \varepsilon) = \gamma_I(\nabla\xi_k, \varepsilon). \tag{761}$$

Recall that the implicit function γ_I appearing in this estimate is one of finitely many different functions $\gamma_I = \gamma_{f(I)}$ from Section (7.3.6) which depend only on the direction coordinate $f(I) \in F$. For the present proof, we will omit the subscript $k = k(I) \in (\mathbb{Z}/2\mathbb{Z})^3 \times \mathbb{Z}$ in the location coordinate for $\nabla\xi = \nabla\xi_k$, so that k may be used to denote the order of differentiation in the bounds of Proposition (20.2) and the calculations below. It is only important to remember that for each index I, only phase gradients ξ_J on the same region $k(J) = k(I)$ enter into (761).

According to the estimates (17.3), (577) and (589) in Section (17), each material derivative on $\nabla\xi$ costs $\Xi e_v^{1/2}$ and each spatial derivative on $\nabla\xi$, $\frac{\bar{D}\nabla\xi}{\partial t}$ or $\frac{\bar{D}^2\nabla\xi}{\partial t^2}$ costs either Ξ or $N^{1/L}\Xi$ depending on whether or not the total order of differentiation on ξ exceeds L.

For spatial derivatives of γ_I, the terms with the highest order derivatives are schematically given by

$$\nabla^k[\gamma_I(\nabla\xi, \varepsilon)] \subseteq [\partial_p \gamma](\nabla^{k+1}\xi) + [\partial_\varepsilon \gamma](\nabla^k \varepsilon) + \text{ cross terms.} \tag{762}$$

As we have seen already in several cases such as Section (18.5), the cross terms are lower order in terms of powers of $N^{1/L}$.

The two main terms in (762) obey the same estimates as each other, except that the $k + 1$ derivative of ξ costs an extra factor of $N^{1/L}$.

$$\|\nabla^{k+1}\xi\|_{C^0} \leq C_k N^{(k+1-L)_+/L} \Xi^k \tag{763}$$

$$\|\nabla^k \varepsilon\|_{C^0} \leq C_k N^{(k-L)_+/L} \Xi^k \tag{764}$$

All the derivatives of the functions γ_I which appear in the expansion are bounded by universal constants, giving (758).

Taking a material derivative of γ_I and differentiating in space gives terms that are schematically of the form

$$\frac{\bar{D}}{\partial t}[\gamma_I(\nabla\xi,\varepsilon)] \subseteq [\partial_p\gamma]\frac{\bar{D}\nabla\xi}{\partial t} + [\partial_\varepsilon\gamma]\frac{\bar{D}\varepsilon}{\partial t} \tag{765}$$

$$\nabla^k\frac{\bar{D}}{\partial t}[\gamma_I(\nabla\xi,\varepsilon)] \subseteq [\partial_p\gamma]\nabla^k\frac{\bar{D}\nabla\xi}{\partial t} + [\partial_p^2\gamma]\nabla^{k+1}\xi\frac{\bar{D}\nabla\xi}{\partial t}$$

$$+ [\partial_\varepsilon\gamma]\nabla^k\frac{\bar{D}\varepsilon}{\partial t} + [\partial_p\partial_\varepsilon\gamma]\nabla^{k+1}\xi\frac{\bar{D}\varepsilon}{\partial t} \tag{766}$$

$$+ \text{ smaller cross terms.}$$

By Proposition (17.3), (577), and (750) all of these terms are bounded by

$$C_k N^{(k+1-L)_+/L}\Xi^{k+1}e_v^{1/2}.$$

In particular, no extra factor of $N^{1/L}$ appears for the first material derivative of $\nabla\xi$; see Section (17) for a review of how this fact follows from the transport equation for $\nabla\xi$.

Finally, we consider

$$\frac{\bar{D}^2}{\partial t^2}[\gamma_I(\nabla\xi,\varepsilon)] \subseteq [\partial_p\gamma]\frac{\bar{D}^2\nabla\xi}{\partial t^2} + [\partial_\varepsilon\gamma]\frac{\bar{D}^2\varepsilon}{\partial t^2} \tag{767}$$

$$+ [\partial_p^2\gamma](\frac{\bar{D}\nabla\xi}{\partial t})^2 + [\partial_p\partial_\varepsilon\gamma](\frac{\bar{D}\nabla\xi}{\partial t})(\frac{\bar{D}\varepsilon}{\partial t}) + [\partial_\varepsilon^2\gamma](\frac{\bar{D}\varepsilon}{\partial t})^2. \tag{768}$$

The terms in Line (768) are lower order, so schematically the main terms from the estimates come from derivatives of the terms in (767), which include

$$\nabla^k\frac{\bar{D}^2}{\partial t^2}[\gamma_I(\nabla\xi,\varepsilon)] \subseteq [\partial_p\gamma]\nabla^k\frac{\bar{D}^2\nabla\xi}{\partial t^2} + [\partial_\varepsilon\gamma]\nabla^k\frac{\bar{D}^2\varepsilon}{\partial t^2} \tag{769}$$

$$+ [\partial_p^2\gamma]\nabla^{k+1}\xi\frac{\bar{D}^2\nabla\xi}{\partial t^2} + [\partial_p\partial_c\gamma]\nabla^{k+1}\xi\frac{\bar{D}^2\varepsilon}{\partial t^2}. \tag{770}$$

At this stage, the second material derivative contributes towards the power of $N^{1/L}$ appearing in the estimates for both the phase gradients and for ε. The only difference is that while the second material derivative costs $\Xi e_v^{1/2}$ when applied to the phase gradients, the second material derivative of ε costs a larger factor of $N\Xi e_R^{1/2}$. Thus, the final bound for (760) is exactly the same quality as the corresponding bound for ε. $\qquad\square$

21 Bounds for the Vector Amplitudes

Having estimated the coefficients γ_I, we are now ready to begin estimating the vector amplitudes v_I and w_I of the correction.

Let us recall once more that v_I is given by $a_I + ib_I$, where

$$b_I^l = \eta\left(\frac{t - t(I)}{\tau}\right)\psi_k(t, x)e^{1/2}(t)\gamma_I P_I^\perp(\nabla\xi_{\sigma I})^l \tag{771}$$

and the real part

$$a_I = -\frac{(\nabla\xi_I)}{|\nabla\xi_I|} \times b_I.$$

is obtained by a $\pi/2$ rotation of b_I in $\langle\nabla\xi_I\rangle^\perp$.

Let us compress the notation slightly by writing

$$v_I^l = \eta\left(\frac{t - t(I)}{\tau}\right)e^{1/2}(t)\gamma_I\alpha_I^l \tag{772}$$

$$\alpha_I^l = [i - \frac{(\nabla\xi_I)}{|\nabla\xi_I|}\times]\psi_k P_I^\perp(\nabla\xi_{\sigma I})^l \tag{773}$$

The vector field α_I^l takes values in the positive eigenspace of the operator $(i\nabla\xi_I)\times$ within the plane $\langle\nabla\xi_I\rangle^\perp$, and is composed entirely from solutions to the transport equation (ψ itself being no worse than ξ), so it obeys the following estimates.

Proposition 21.1 (Estimates for the Transport Part of v_I).

$$\|\nabla^k\alpha_I^l\|_{C^0} \leq C_k N^{(k+1-L)_+/L}\Xi^k \tag{774}$$

$$\|\nabla^k\frac{\bar{D}}{\partial t}\alpha_I^l\|_{C^0} \leq C_k N^{(k+1-L)_+/L}\Xi^{k+1}e_v^{1/2} \tag{775}$$

$$\|\nabla^k\frac{\bar{D}^2}{\partial t^2}\alpha_I^l\|_{C^0} \leq C_k N^{(k+2-L)_+/L}\Xi^{k+2}e_v \tag{776}$$

or more concisely, using the notation of Section (17.3)

$$\|D^{(k,r)}\alpha_I^l\|_{C^0} \leq C_{k+r}N^{((r-1)_+ + k+1-L)_+/L}\Xi^{k+r}e_v^{r/2}. \tag{777}$$

The correction itself is composed of terms

$$V_I = \nabla \times W_I$$

with

$$W_I = \frac{1}{\lambda}e^{i\lambda\xi_I}w_I \tag{778}$$

$$w_I = \frac{v_I}{|\nabla\xi_I|} \tag{779}$$

so w_I also takes the form (772), with a slightly different $\alpha_I^l(w)$ that obeys the same bounds.

Let us now compute estimates for v_I, starting with the spatial derivatives.

139

Proposition 21.2 (Spatial Derivatives of Vector Amplitudes).

$$\|\nabla^k v_I\|_{C^0} + \|\nabla^k w_I\|_{C^0} \leq C_k N^{(k+1-L)_+/L} \Xi^k e_R^{1/2} \tag{780}$$

Proof. Let ∇^k be any k'th order spatial derivative. Then, the main terms from differentiating (772) are

$$\nabla^k v_I^l \subseteq \eta_I(t) e^{1/2}(t) [\nabla^k \gamma_I \alpha_I^l + \gamma_I \nabla^k \alpha_I^l]. \tag{781}$$

The estimate (780) follows from (774) and (758), where the factor $e_R^{1/2}$ comes from the bound (250) on $e^{1/2}(t)$. $\qquad\qquad\square$

Let us now study the first material derivative $\frac{\bar{D}}{\partial t} v_I^l$. It takes on the form

$$\frac{\bar{D} v_I^l}{\partial t} = \tau^{-1} \eta' \left(\frac{t - t(I)}{\tau} \right) e^{1/2}(t) \gamma_I \alpha_I^l + \eta_I(t) \frac{de^{1/2}}{dt} \gamma_I \alpha_I^l \tag{782}$$

$$+ \eta_I(t) e^{1/2}(t) \left(\frac{\bar{D} \gamma_I}{\partial t} \alpha_I^l + \gamma_I \frac{\bar{D} \alpha_I^l}{\partial t} \right). \tag{783}$$

For every term besides the time cutoff, the first material derivative costs $\Xi e_v^{1/2}$, whereas for the time cutoff, we lose a factor

$$\tau^{-1} = b^{-1} \Xi e_v^{1/2} = b_0^{-1} B_\lambda^{1/2} \left(\frac{e_R^{1/2} N}{e_v^{1/2}} \right)^{1/2}.$$

This loss leads to the following bounds.

Proposition 21.3 (First Material Derivative of Vector Amplitudes).

$$\|\nabla^k \frac{\bar{D} v_I^l}{\partial t}\|_{C^0} + \|\nabla^k \frac{\bar{D} w_I^l}{\partial t}\|_{C^0}$$

$$\leq C_k B_\lambda^{1/2} \left(\frac{e_R^{1/2} N}{e_v^{1/2}} \right)^{1/2} N^{(k+1-L)_+} \Xi^{k+1} e_v^{1/2} e_R^{1/2}$$

For the second material derivative, there are two factors which lose more than a factor of $\Xi e_v^{1/2}$: differentiating the time cutoff again gives another factor of τ^{-1}, whereas the second material derivative of γ_I gives rise to a factor of ϵ_t^{-1}. It turns out that these extra factors are in balance with each other; namely

$$\tau^{-2} \leq C \left[B_\lambda^{1/2} \left(\frac{e_R^{1/2} N}{e_v^{1/2}} \right)^{1/2} \right]^2 (\Xi e_v^{1/2})^2 \tag{784}$$

$$\leq C B_\lambda N \Xi e_v^{1/2} e_R^{1/2} \tag{785}$$

$$\epsilon_t^{-1} (\Xi e_v^{1/2}) \leq C (N e_R^{1/2} \Xi) (\Xi e_v^{1/2}) \tag{786}$$

$$\leq C N \Xi e_v^{1/2} e_R^{1/2}. \tag{787}$$

This observation leads to the following bounds.

Proposition 21.4 (The Second Material Derivative of the Vector Amplitude).

$$\|\nabla^k \frac{\bar{D}^2 v_I}{\partial t^2}\|_{C^0} + \|\nabla^k \frac{\bar{D}^2 w_I}{\partial t^2}\|_{C^0} \le C_k B_\lambda N^{1+(k+1-L)_+/L} \Xi^{k+2} e_v^{1/2} e_R \quad (788)$$

Proof. Let us write the second material derivative of v_I schematically as

$$\frac{\bar{D}^2 v_I}{\partial t^2} = \frac{d^2}{dt^2} [\eta \left(\frac{t - t(I)}{\tau} \right) e^{1/2}(t)] \gamma_I \alpha_I^l + \eta_I(t) e^{1/2}(t) \frac{\bar{D}^2}{\partial t^2} [\gamma_I \alpha_I^l]$$

$$+ \text{ smaller cross terms}$$

$$= A_{(I)} + A_{(II)}.$$

Now, by (774) and (758), the first of these terms is bounded by

$$\|\nabla^k A_{(I)}\|_{C^0} \le C \tau^{-2} N^{(k+1-L)_+/L} \Xi^k e_R^{1/2} \quad (789)$$

$$\le C b^{-2} N^{(k+1-L)_+/L} \Xi^{k+2} e_v e_R^{1/2} \quad (790)$$

$$\le C B_\lambda \left(\frac{e_R^{1/2} N}{e_v^{1/2}} \right) N^{(k+1-L)_+/L} \Xi^{k+2} e_v e_R^{1/2} \quad (791)$$

$$\le C B_\lambda N^{1+(k+1-L)_+/L} \Xi^{k+2} e_v^{1/2} e_R. \quad (792)$$

For the second of these terms

$$A_{(II)} = \frac{\bar{D}^2}{\partial t^2} [\gamma_I \alpha_I^l] \quad (793)$$

we compare

$$\nabla^k A_{(II)} \subseteq \nabla^k \frac{\bar{D}^2 \gamma_I}{\partial t^2} \alpha_I^l + \gamma_I \nabla^k \frac{\bar{D}^2 \alpha_I^l}{\partial t^2}. \quad (794)$$

All the bounds for γ_I and α_I are identical until the second material derivative. For the second material derivative, we use (760) and (776) to compare

$$\|\nabla^k \frac{\bar{D}^2 \gamma_I}{\partial t^2}\|_{C^0} \le C_k N^{1+(k+1-L)_+/L} \Xi^{k+2} e_v^{1/2} e_R^{1/2} \quad (795)$$

$$\|\nabla^k \frac{\bar{D}^2 \alpha_I}{\partial t^2}\|_{C^0} \le C_k N^{(k+2-L)_+/L} \Xi^{k+2} e_v. \quad (796)$$

In fact, the bound for γ_I is the larger of the two, as can be seen by proving

$$N^{(k+2-L)_+/L} e_v^{1/2} \le N^{1+(k+1-L)_+/L} e_R^{1/2} \quad (797)$$

$$\Leftrightarrow \left(\frac{e_v}{e_R} \right)^{1/2} \le N^{1+[(k+1-L)_+ - (k+2-L)_+/L]} \quad (798)$$

but

$$N^{1+[(k+1-L)_+ - (k+2-L)_+/L]} \ge N^{1-1/L} \ge N^{1/2}$$

141

so (798) follows from the fact that

$$N \geq \left(\frac{e_v}{e_R}\right)^{3/2} \geq \left(\frac{e_v}{e_R}\right).$$

Also observe that the bound for $\|\nabla^k \frac{\bar{D}^2 \gamma_I}{\partial t^2}\|_{C^0}$ is the same bound as for $\|\nabla^k A_{(I)}\|_{C^0}$ without the large constant B_λ, which concludes the proof of (21.4).

$\square$

The results of this section also extend to the corrected amplitude

$$\tilde{v}_I = v_I + \frac{\nabla \times w_I}{B_\lambda N \Xi}$$

appearing in the representation

$$V_I = e^{i\lambda \xi_I} \tilde{v}_I^l.$$

Corollary 21.1 (Estimates for Corrected Amplitudes). *The vector field*

$$\tilde{v}_I = v_I + \frac{\nabla \times w_I}{B_\lambda N \Xi}$$

also satisfies all the estimates stated in Section (21) for v_I.

Proof. The bounds for $\|\nabla^k v_I\|_{C^0}$ are clear since

$$\nabla^k \tilde{v}_I = \nabla^k v_I + \frac{1}{B_\lambda N \Xi} \nabla^k \nabla \times w_I \tag{799}$$

$$\|\nabla^k \tilde{v}_I\|_{C^0} \leq C_k N^{(k+1-L)_+/L} \Xi^k e_R^{1/2} \\ + (B_\lambda N \Xi)^{-1} \cdot (N^{(k+2-L)_+/L} \Xi^{k+1} e_R^{1/2}) \tag{800}$$

$$\leq C_k N^{(k+1-L)_+/L} \Xi^k e_R^{1/2}. \tag{801}$$

To estimate $\frac{\bar{D}\tilde{v}_I}{\partial t}$ we write

$$\frac{\bar{D}\tilde{v}_I}{\partial t} = \frac{\bar{D}v_I}{\partial t} + \frac{1}{B_\lambda N \Xi}[\nabla \times (\frac{\bar{D}w_I}{\partial t}) + D[v_\epsilon](w)] \tag{802}$$

where the commutator term is an operator which is schematically of the form

$$D[v_\epsilon](w) = \nabla v_\epsilon \cdot \nabla w. \tag{803}$$

From this expression, it is clear that the commutator gives a cost of $\Xi e_v^{1/2}$, whereas the material derivative itself carries a larger cost of $\tau^{-1} = b^{-1}\Xi e_v^{1/2}$ in the estimates.

$\square$

22 Bounds for the Corrections

We now give bounds for the correction terms using the estimates of the preceding sections.

22.1 Bounds for the Velocity Correction

Since V is of the form

$$V = \nabla \times W$$

estimating V and its derivatives will follow from estimating the derivatives of

$$W = \sum_I W_I \tag{804}$$

$$W_I = (B_\lambda N\Xi)^{-1} e^{i\lambda \xi_I} w_I. \tag{805}$$

Let us first estimate the spatial derivatives.

Proposition 22.1 (Spatial Derivatives of W).

$$\|\nabla^k W\|_{C^0} \leq C_k (B_\lambda N\Xi)^{k-1} e_R^{1/2} \tag{806}$$

Proof. Since the number of W_I supported at any given region of $\mathbb{R} \times \mathbb{T}^3$ is bounded by a universal constant, it suffices to estimate W_I uniformly in I. For an individual wave, it is easy to see that the estimate will hold. At the level of $\|W\|_{C^0}$ we have

$$\|(B_\lambda N\Xi)^{-1} e^{i\lambda \xi_I} w_I\|_{C^0} \leq C(B_\lambda N\Xi)^{-1} e_R^{1/2}. \tag{807}$$

For the derivatives, we apply the product rule to

$$\nabla^k e^{i\lambda \xi_I} w_I = \sum_{|a|+|b|=k} \nabla^a [e^{i\lambda \xi_I}] \nabla^b [w_I]. \tag{808}$$

During repeated differentiation, the derivative hits either

- the oscillatory factor $e^{i\lambda \xi_I}$, which costs $C\lambda = CB_\lambda N\Xi$;

- the phase direction $\nabla \xi_I$ or one of its derivatives, which costs at most $CN^{1/L}\Xi$; or

- the amplitude w_I or one of its derivatives, which costs at most a factor of $CN^{1/L}\Xi$.

In any case, the largest cost happens when differentiating the phase function, leading to the estimate in Proposition (22.1). $\square$

We can similarly give estimates for derivatives of the coarse scale material derivative of W.

Proposition 22.2 (Coarse Scale Material Derivative of W).

$$\|\nabla^k \frac{\bar{D}W}{\partial t}\|_{C^0} \le C_k B_\lambda^{1/2} \left(\frac{e_R^{1/2} N}{e_v^{1/2}} \right)^{1/2} (B_\lambda N\Xi)^{k-1} \Xi e_v^{1/2} e_R^{1/2} \qquad (809)$$

Proof. The proof is by the same method as in Proposition (22.1). Here we simply observe that because the phase is transported by the coarse scale flow

$$\frac{\bar{D}W_I}{\partial t} = \frac{1}{\lambda} e^{i\lambda \xi_I} \frac{\bar{D}w_I}{\partial t} \qquad (810)$$

where the amplitude of $\frac{\bar{D}w_I}{\partial t}$ is bounded by

$$\left\| \frac{\bar{D}w_I}{\partial t} \right\|_{C^0} \le C\tau^{-1} e_R^{1/2} \le CB_\lambda^{1/2} \left(\frac{e_R^{1/2} N}{e_v^{1/2}} \right)^{1/2} \Xi e_v^{1/2} e_R^{1/2}.$$

Upon taking spatial derivatives, the largest cost of $B_\lambda N\Xi$ always occurs when the derivative hits the oscillatory factor. $\qquad \square$

The Main Lemma (10.1) asks for bounds on

$$(\partial_t + v \cdot \nabla)W$$

rather than

$$\frac{\bar{D}W}{\partial t} = (\partial_t + v_\epsilon \cdot \nabla)W.$$

As the following lemma illustrates, it is always possible to obtain estimates for $(\partial_t + v \cdot \nabla)$ of a quantity, once one has appropriate estimates for coarse scale material derivatives and for spatial derivatives.

Corollary 22.1 (Estimates for $(\partial_t + v \cdot \nabla)W$). *For all $k = 0, \ldots, L$, the bounds for $\nabla^k \frac{\bar{D}W}{\partial t}$ and $\nabla^k (\partial_t + v \cdot \nabla)W$ are of the same order. More precisely,*

$$\|\nabla^k (\partial_t + v \cdot \nabla)W\|_{C^0} \le C_k B_\lambda^{1/2} \left(\frac{e_R^{1/2} N}{e_v^{1/2}} \right)^{1/2} (B_\lambda N\Xi)^{k-1} \Xi e_v^{1/2} e_R^{1/2}. \qquad (811)$$

Proof. For $k = 0$, we simply write

$$(\partial_t + v \cdot \nabla) = (\partial_t + v_\epsilon \cdot \nabla) + (v - v_\epsilon) \cdot \nabla.$$

It then suffices to estimate

$$(v - v_\epsilon) \cdot \nabla W = (v^a - v_\epsilon^a)\partial_a W \qquad (812)$$

and its derivatives. At the level of C^0, we have

$$\|v - v_\epsilon\|_{C^0} \leq C\frac{e_v^{1/2}}{N} \tag{813}$$

by (470).

Since a spatial derivative of W costs $B_\lambda N\Xi$, we see that the cost of $(v - v_\epsilon) \cdot \nabla$ is at most $\Xi e_v^{1/2}$.

$$\|(v - v_\epsilon) \cdot \nabla W\|_{C^0} \leq C[\Xi e_v^{1/2}](B_\lambda N\Xi)^{-1}e_R^{1/2} \tag{814}$$

which is better than the cost of $\tau^{-1} = b^{-1}\Xi e_v^{1/2}$ of $\frac{\bar{D}}{\partial t}$.

For the spatial derivatives, we are allowed to lose a factor of $N\Xi$ per spatial derivative. Comparing the bounds

$$\|v - v_\epsilon\|_{C^0} \leq C\frac{e_v^{1/2}}{N} \tag{815}$$

$$\|\nabla v\|_{C^0} + \|\nabla v_\epsilon\|_{C^0} \leq \Xi e_v^{1/2} \tag{816}$$

we see that our estimate worsens by a factor of $N\Xi$ when we treat the terms ∇v and ∇v_ϵ separately. We can now proceed to take up to L derivatives of v altogether, giving the corollary. $\quad\square$

Corollary 22.2 (Spatial Derivatives of V).

$$\|\nabla^k V\|_{C^0} \leq C_k(B_\lambda N\Xi)^k e_R^{1/2} \tag{817}$$

Proof. This follows from (22.1) and the fact that $V = \nabla \times W$. $\quad\square$

Corollary 22.3 (Coarse Scale Material Derivative of V).

$$\|\nabla^k\frac{\bar{D}V}{\partial t}\|_{C^0} \leq C_k B_\lambda^{1/2}\left(\frac{e_R^{1/2}N}{e_v^{1/2}}\right)^{1/2}(B_\lambda N\Xi)^k\Xi e_v^{1/2}e_R^{1/2} \tag{818}$$

Proof. This estimate follows from (22.2) and (22.1) after we express

$$\frac{\bar{D}}{\partial t}V = \frac{\bar{D}}{\partial t}\nabla \times W \tag{819}$$

$$= \nabla \times \frac{\bar{D}W}{\partial t} - D[v_\epsilon][W] \tag{820}$$

where the commutator $D[v_\epsilon][W]$ is a spatial derivative operator of the form

$$D[v_\epsilon][W] \approx \nabla v_\epsilon \nabla W. \tag{821}$$

The material derivative costs $\tau^{-1} = b^{-1}\Xi e_v^{1/2}$

$$\|\nabla \times \frac{\bar{D}W}{\partial t}\|_{C^0} \leq C(B_\lambda N\Xi)[(B_\lambda N\Xi)^{-1}\tau^{-1}e_R^{1/2}] \tag{822}$$

145

whereas the commutator only costs $(\Xi e_v^{1/2})$

$$\|\nabla v_\epsilon \nabla W\|_{C^0} \le C(B_\lambda N\Xi)[(B_\lambda N\Xi)^{-1}(\Xi e_v^{1/2})e_R^{1/2}]. \tag{823}$$

Each additional spatial derivative costs a factor $(B_\lambda N\Xi)$ as it falls on the oscillatory factor in W. □

Corollary 22.4 (Estimates for $\nabla^k(\partial_t + v \cdot \nabla)V$). *For all* $k = 0, \ldots, L$, *the bounds for* $\nabla^k \frac{\bar{D}V}{\partial t}$ *and* $\nabla^k(\partial_t + v \cdot \nabla)V$ *are of the same order. More precisely,*

$$\|\nabla^k(\partial_t + v \cdot \nabla)V\|_{C^0} \le C_k B_\lambda^{1/2} \left(\frac{e_R^{1/2}N}{e_v^{1/2}}\right)^{1/2} (B_\lambda N\Xi)^k \Xi e_v^{1/2} e_R^{1/2}. \tag{824}$$

Proof. The proof is identical to the proof of Corollary (22.1). □

22.2 Bounds for the Pressure Correction

The correction P to the pressure consists of two parts.

$$P = P_0 + \sum_{J \ne \bar{I}} P_{I,J} \tag{825}$$

$$P_0 = -\frac{e(t)}{3} - \frac{R_\epsilon^{jl}\delta_{jl}}{3} \tag{826}$$

$$P_{I,J} = -\frac{V_I \cdot V_J}{2} \tag{827}$$

$$= -\frac{1}{2}e^{i\lambda(\xi_I + \xi_J)}\tilde{v}_I \cdot \tilde{v}_J \tag{828}$$

From the preceding section, we can see that the estimates for the high frequency part $V_I \cdot V_J$ are the dominant ones, since we have seen that the primary cost of spatial derivatives comes from the oscillatory factor, and the primary cost of material derivatives comes from the time cutoff.

We also see from the quadratic nature of formula (827) that the estimates for $P_{I,J}$ will gain a factor $e_R^{1/2}$ compared to those of V_I. These estimates are summarized by the following proposition.

Proposition 22.3 (Estimates for Pressure Correction).

$$\|\nabla^k P\|_{C^0} \le C_k(B_\lambda N\Xi)^k e_R \tag{829}$$

$$\|\nabla^k \frac{\bar{D}P}{\partial t}\|_{C^0} \le C_k B_\lambda^{1/2} \left(\frac{e_R^{1/2}N}{e_v^{1/2}}\right)^{1/2} (B_\lambda N\Xi)^k \Xi e_v^{1/2} e_R \tag{830}$$

Also, for $k = 0, \ldots, L$, *we have*

$$\|\nabla^k(\partial_t + v \cdot \nabla)P\|_{C^0} \le C_k B_\lambda^{1/2} \left(\frac{e_R^{1/2}N}{e_v^{1/2}}\right)^{1/2} (B_\lambda N\Xi)^k \Xi e_v^{1/2} e_R. \tag{831}$$

For example, by Propositions (22.2) and (18.3)

$$\|P_0\|_{C^0} \leq C(\|e(t)\|_{C^0(\mathbb{R})} + \|R_\epsilon\|_{C^0}) \leq Ce_R \tag{832}$$

$$\|P_{I,J}\|_{C^0} \leq \frac{1}{2}(\|V_I\|_{C^0}\|V_J\|_{C^0}) \leq Ce_R \tag{833}$$

$$\|\nabla P_0\|_{C^0} \leq C\|\nabla R_\epsilon\|_{C^0} \leq C\Xi e_R \tag{834}$$

$$\|\nabla P_{I,J}\|_{C^0} \leq \frac{1}{2}(\|\nabla V_I\|_{C^0}\|V_J\|_{C^0} + \|V_I\|_{C^0}\|\nabla V_J\|_{C^0}) \tag{835}$$

$$\leq C(N\Xi e_R^{1/2})e_R^{1/2} = CN\Xi e_R. \tag{836}$$

For the first material derivative of the low frequency term, we can apply (813) and Proposition (18.4).

$$\|(\partial_t + v \cdot \nabla)P_0\|_{C^0} = \|\frac{\bar{D}P_0}{\partial t} + (v - v_\epsilon) \cdot \nabla P_0\|_{C^0} \tag{837}$$

$$\leq \|\frac{\bar{D}P_0}{\partial t}\|_{C^0} + \|v - v_\epsilon\|_{C^0}\|\nabla P_0\|_{C^0} \tag{838}$$

$$\|v - v_\epsilon\|_{C^0}\|\nabla P_0\|_{C^0} \leq (\frac{e_v^{1/2}}{N})(\Xi e_R) \tag{839}$$

$$\|\frac{\bar{D}P_0}{\partial t}\|_{C^0} \leq C(\|e'(t)\|_{C^0(\mathbb{R})} + \|\frac{\bar{D}R_\epsilon}{\partial t}\|_{C^0}) \tag{840}$$

$$\leq C\Xi e_v^{1/2}e_R \tag{841}$$

$$\|(\partial_t + v \cdot \nabla)P_0\|_{C^0} \leq C\Xi e_v^{1/2}e_R \tag{842}$$

For the first material derivative of the high frequency terms, we have

$$\|(\partial_t + v \cdot \nabla)P_{I,J}\|_{C^0} \leq \|\frac{\bar{D}P_{I,J}}{\partial t}\|_{C^0} + \|v - v_\epsilon\|_{C^0}\|\nabla P_{I,J}\|_{C^0} \tag{843}$$

$$\|v - v_\epsilon\|_{C^0}\|\nabla P_{I,J}\|_{C^0} \leq C(\frac{e_v^{1/2}}{N})(N\Xi e_R) \tag{844}$$

$$\leq C\Xi e_v^{1/2}e_R \tag{845}$$

$$\|\frac{\bar{D}P_{I,J}}{\partial t}\|_{C^0} \leq (\|\frac{\bar{D}V_I}{\partial t}\|_{C^0}\|V_J\|_{C^0} + \|V_I\|_{C^0}\|\frac{\bar{D}V_J}{\partial t}\|_{C^0}) \tag{846}$$

$$\leq CB_\lambda^{1/2}\left(\frac{e_R^{1/2}N}{e_v^{1/2}}\right)^{1/2}\Xi e_v^{1/2}e_R^{1/2}. \tag{847}$$

The bounds stated for $\|\frac{\bar{D}V_I}{\partial t}\|_{C^0}$ in the preceding inequality do not follow directly from Corollary (22.4), but follow from the same proof.

23 Energy Approximation

In this section, we will establish the estimates (261) and (262) of the Main Lemma (10.1), which state that we are able to accurately prescribe the energy

$$\int |V|^2(t,x)dx \approx \int_{\mathbb{T}^3} e(t)dx \tag{848}$$

that the correction adds to the solution, and also bound the difference between the time derivatives of these two quantities.

Quantitatively, we have the following bound.

Proposition 23.1 (Prescribing Energy; Proof of (261)).

$$\left\| \int |V|^2(t,x)dx - \int_{\mathbb{T}^3} e(t)dx \right\|_{C_t^0} \le C\frac{e_R}{N} \tag{849}$$

Proof. We calculate

$$\int |V|^2(t,x)dx = \int V^j V^l \delta_{jl} dx$$

$$= \int \sum_{I,J} V_I^j V_J^l \delta_{jl} dx$$

$$= \sum_I \int V_I^j \bar{V}_I^l \delta_{jl} dx + \sum_{J \ne \bar{I}} \int V_I \cdot V_J dx$$

$$= \sum_I \int \tilde{v}_I^j \bar{\tilde{v}}_I^l \delta_{jl} dx + \sum_{J \ne \bar{I}} \int e^{i\lambda(\xi_I + \xi_J)} \tilde{v}_I \cdot \tilde{v}_J dx$$

$$= \int \left(\sum_I v_I^j \bar{v}_I^l \right) \delta_{jl} dx + \sum_I \int \frac{(\nabla \times w_I)}{\lambda} \cdot \bar{v}_I dx$$

$$+ \sum_I \int \frac{(\nabla \times \bar{w}_I)}{\lambda} \cdot v_I dx + \sum_I \int \frac{(\nabla \times w_I)}{\lambda} \cdot \frac{(\nabla \times w_I)}{\lambda} dx$$

$$+ \sum_{J \ne \bar{I}} \int e^{i\lambda(\xi_I + \xi_J)} \tilde{v}_I \cdot \tilde{v}_J dx$$

$$= \int e(t)dx + E_{(A)}(t) + \sum_{J \ne \bar{I}} E_{(B,IJ)}(t). \tag{850}$$

where it is clear that we have a bound for

$$|E_{(A)}(t)| \le 2\sum_I \int \left| \frac{(\nabla \times w_I)}{\lambda} \cdot \bar{v}_I \right| dx$$

$$+ \sum_I \int \left| \frac{(\nabla \times w_I)}{\lambda} \cdot \frac{(\nabla \times w_I)}{\lambda} \right| dx \tag{851}$$

$$\le C\left(\frac{\Xi e_R^{1/2} \cdot e_R^{1/2}}{B_\lambda \Xi N} + \left(\frac{\Xi e_R^{1/2}}{B_\lambda \Xi N} \right)^2 \right) \tag{852}$$

$$\le C\frac{e_R}{N} \tag{853}$$

and each integral contributing to the other term can be estimated by integrating by parts

$$E_{(B,IJ)}(t) = \int e^{i\lambda(\xi_I + \xi_J)} \tilde{v}_I \cdot \tilde{v}_J dx \tag{854}$$

$$= \int \left[\frac{\partial^a (\xi_I + \xi_J)}{i\lambda |\nabla(\xi_I + \xi_J)|^2} \partial_a [e^{i\lambda(\xi_I + \xi_J)}] \right] \tilde{v}_I \cdot \tilde{v}_J dx \tag{855}$$

$$= \frac{-1}{i\lambda} \int e^{i\lambda(\xi_I + \xi_J)} \partial_a \left[\frac{\partial^a (\xi_I + \xi_J)}{|\nabla(\xi_I + \xi_J)|^2} \tilde{v}_I \cdot \tilde{v}_J \right] dx \tag{856}$$

$$|E_{(B,IJ)}(t)| \leq C \frac{\Xi e_R}{B_\lambda N \Xi} \tag{857}$$

$$\square$$

Similarly, we can estimate our control over the rate of energy variation.

Proposition 23.2 (Rate of Energy Variation Estimate; Proof of (262)).

$$\left\| \frac{d}{dt} \left[\int |V|^2(t,x) dx - \int_{\mathbb{T}^3} e(t) dx \right] \right\|_{C_t^0} \leq C B_\lambda^{1/2} \left(\frac{e_v^{1/2}}{e_R^{1/2} N} \right)^{-1/2} (\Xi e_v^{1/2}) \frac{e_R}{N} \tag{858}$$

Proof. We again use the expression

$$\int |V|^2(t,x) dx - \int e(t) dx = E_{(A)}(t) + E_{(B)}(t) \tag{859}$$

$$E_{(A)}(t) = \sum_I \int \frac{(\nabla \times w_I)}{\lambda} \cdot \bar{v}_I dx + \sum_I \int \frac{(\nabla \times \bar{w}_I)}{\lambda} \cdot v_I dx \tag{860}$$

$$+ \sum_I \int \frac{(\nabla \times w_I)}{\lambda} \cdot \frac{(\nabla \times w_I)}{\lambda} dx \tag{861}$$

$$E_{(B)}(t) = \sum_{J \neq \bar{I}} \int e^{i\lambda(\xi_I + \xi_J)} \tilde{v}_I \cdot \tilde{v}_J dx \tag{862}$$

and differentiate in time, starting with

$$\frac{d}{dt} E_{(A)}(t)$$
$$= \sum_I \int \frac{\bar{D}}{\partial t} \left[\frac{(\nabla \times w_I)}{\lambda} \cdot \bar{v}_I + \frac{(\nabla \times \bar{w}_I)}{\lambda} \cdot v_I + \frac{(\nabla \times w_I)}{\lambda} \cdot \frac{(\nabla \times w_I)}{\lambda} \right] dx. \tag{863}$$

As we know, the coarse scale material derivative $\frac{\bar{D}}{\partial t}$ costs a factor

$$|\frac{\bar{D}}{\partial t}| \leq C B_\lambda^{1/2} \left(\frac{e_v^{1/2}}{e_R^{1/2} N} \right)^{-1/2} \Xi e_v^{1/2}$$

giving

$$|\frac{d}{dt}E_{(A)}(t)| \le CB_\lambda^{1/2}\left(\frac{e_v^{1/2}}{e_R^{1/2}N}\right)^{-1/2}\Xi e_v^{1/2}\frac{e_R}{N} \tag{864}$$

and similarly, we can write

$$\frac{d}{dt}E_{(B)}(t) = \sum_{J\ne \bar{I}}\int \frac{\bar{D}}{\partial t}\left[e^{i\lambda(\xi_I+\xi_J)}\tilde{v}_I \cdot \tilde{v}_J\right]dx \tag{865}$$

$$= \sum_{J\ne \bar{I}}\int e^{i\lambda(\xi_I+\xi_J)}\frac{\bar{D}}{\partial t}\left[\tilde{v}_I \cdot \tilde{v}_J\right]dx, \tag{866}$$

which can be estimated by

$$|\frac{d}{dt}E_{(B)}(t)| \le CB_\lambda^{1/2}\left(\frac{e_v^{1/2}}{e_R^{1/2}N}\right)^{-1/2}\Xi e_v^{1/2}\frac{e_R}{N} \tag{867}$$

using another integration by parts, just as in (856). $\qquad\square$

24 Checking Frequency Energy Levels for the Velocity and Pressure

Since we have established many estimates for the corrections to the velocity and the pressure, we are in a position to compare with the Main Lemma.

We first remark that the estimates (258)–(260) for W were established in Proposition (22.1) and Corollary (22.1). Similarly, the estimates (255)–(257) were special cases of the Proposition (22.2) and Corollary (22.4). Also, the estimates (263)–(265) were established by Proposition (22.3). At least, all these bounds will be established for a particular constant C once the constant B_λ has been chosen.

We are also in a position to check that the frequency and energy levels of the new velocity and pressure

$$v_1 = v + V \tag{868}$$
$$p_1 = p + P \tag{869}$$

are consistent with the claims of the Main Lemma (10.1).

For the velocity, at the level of the first derivative, we have

$$\|\nabla v_1\|_{C^0} \le \|\nabla v\|_{C^0} + \|\nabla V\|_{C^0} \tag{870}$$

$$\le C(\Xi e_v^{1/2} + NB_\lambda \Xi e_R^{1/2}) \tag{871}$$

$$\le CN\Xi e_R^{1/2} \tag{872}$$

since $N \geq \left(\frac{e_v}{e_R}\right)^{1/2}$ and since B_λ will be chosen to be some constant which has been absorbed into the C.

We also have

$$\|\nabla^k v_1\|_{C^0} \leq C_k (B_\lambda N \Xi)^k e_R^{1/2} \tag{873}$$

for $k = 1, \ldots, L$, which is exactly consistent with the Definition (241) for the new frequency energy levels $\Xi' = CN\Xi$ and $e'_v = e_R$.

For the pressure, we have

$$\|\nabla p_1\|_{C^0} \leq \|\nabla p\|_{C^0} + \|\nabla P\|_{C^0} \tag{874}$$
$$\leq C(\Xi e_v + N B_\lambda \Xi e_R) \tag{875}$$
$$\leq C B_\lambda N \Xi e_R \tag{876}$$

since $N \geq \left(\frac{e_v}{e_R}\right)$. This bound is also exactly consistent with the Definition (242) for the new frequency energy levels $\Xi' = CN\Xi$ and $e'_v = e_R$.

To complete the proof of the Main Lemma (10.1), it now only remains to choose a constant B_λ so that (243) and (244) can be verified for the new energy levels. This choice of B_λ is the last step of the proof.

Part VII
Construction of Regular Weak Solutions: Estimating the New Stress

To conclude the argument, we must calculate the new stress R_1 and verify that this stress obeys all the estimates required in Lemma (10.1).

A complete list of the terms contributing to the new stress R_1 includes the following:

1. **The Mollification Terms**

$$Q_{M,v}^{jl} = (v^j - v_\epsilon^j)V^l + V^j(v^l - v_\epsilon^l) \tag{877}$$

$$Q_{M,R}^{jl} = R^{jl} - R_\epsilon^{jl} \tag{878}$$

2. **The Stress Term**

$$Q_S^{jl} = \sum_I \frac{(\nabla \times w_I)^j \bar{v}_I^l}{\lambda} + \sum_I \frac{v_I^j (\nabla \times \bar{w}_I)^l}{\lambda} \\ + \sum_I \frac{(\nabla \times w_I)^j (\nabla \times \bar{w}_I)^l}{\lambda^2} \tag{879}$$

3. **The High-Low Interaction Term**

$$\partial_j Q_L^{jl} = \sum_I e^{i\lambda \xi_I}(\tilde{v}_I^j \partial_j v_\epsilon^l) \tag{880}$$

4. **The High-High Interference Terms**

$$\partial_j Q_H^{jl} = -\sum_{J \neq \bar{I}} \lambda e^{i\lambda(\xi_I + \xi_J)}[v_I \times (|\nabla \xi_J| - 1)v_J + v_J \times (|\nabla \xi_I| - 1)v_I]$$

$$+ \text{ lower order terms} \tag{881}$$

5. **The Transport Term**

$$\partial_j Q_T^{jl} = \sum_I e^{i\lambda \xi_I}[(\partial_t + v_\epsilon^j \partial_j)\tilde{v}_I^l] \tag{882}$$

For all of these terms, we will verify[18] the following estimates:

$$\|R_1\|_{C^0} \trianglelefteq \frac{1}{10} e_R'$$

$$\Leftrightarrow \qquad \|R_1\|_{C^0} \trianglelefteq \frac{1}{10} \left(\frac{e_v^{1/2}}{e_R^{1/2} N} \right)^{-1/2} \frac{e_v^{1/2} e_R^{1/2}}{N} \tag{883}$$

$$\|\nabla^k R_1\|_{C^0} \trianglelefteq (\Xi')^k e_R' \qquad\qquad k = 1, \ldots, L$$

$$\Leftrightarrow \qquad \|\nabla^k R_1\|_{C^0} \trianglelefteq C(N\Xi)^k \left[\left(\frac{e_v^{1/2}}{e_R^{1/2} N} \right)^{-1/2} \frac{e_v^{1/2} e_R^{1/2}}{N} \right] \tag{884}$$

for $k = 1, \ldots, L$, and we must also verify

$$\|(\partial_t + v_1 \cdot \nabla)R_1\|_{C^0} \trianglelefteq (\Xi'(e_v')^{1/2}) e_R'$$

$$\Leftrightarrow \qquad \|(\partial_t + v_1 \cdot \nabla)R_1\|_{C^0} \trianglelefteq C \left(\frac{e_v^{1/2}}{e_R^{1/2} N} \right)^{-1/2} \Xi e_v^{1/2} e_R \tag{885}$$

$$\|\nabla^k(\partial_t + v_1 \cdot \nabla)R_1\|_{C^0} \trianglelefteq C(N\Xi)^k \left[\left(\frac{e_v^{1/2}}{e_R^{1/2} N} \right)^{-1/2} \Xi e_v^{1/2} e_R \right] \tag{886}$$

$$v_1 - v + V$$

for all $k = 0, \ldots, L - 1$.

Note that, in contrast to the goals (884)–(886), the goal (883) does not allow for an implied constant C. When verifying the goals (884)–(886), the constant C will depend on our choice of B_λ.

The goals (885) and (886) have a slightly inconvenient feature in that while most of our bounds are stated for the coarse scale material derivative

$$\frac{\bar{D}}{\partial t} = (\partial_t + v_\epsilon \cdot \nabla)$$

the goals (885) and (886) require estimates for the derivative

$$(\partial_t + v_1 \cdot \nabla) = (\partial_t + v \cdot \nabla) + V \cdot \nabla \tag{887}$$
$$= (\partial_t + v_\epsilon \cdot \nabla) + [(v - v_\epsilon) + V] \cdot \nabla. \tag{888}$$

Therefore, before we proceed, it is useful to remark that once the spatial derivative bounds (884) have been verified, then the bounds (885) and (886) need only be checked for the derivative $\frac{\bar{D}}{\partial t}$ in place of $(\partial_t + v_1 \cdot \nabla)$. We summarize this observation through the following proposition.

---

[18]Recall that the notation $\trianglelefteq$ refers to inequalities that are goals but that generally have not yet been proven.

Proposition 24.1 (Material Derivative Criterion 1). *There exists a constant C such that for all symmetric $(2,0)$ tensor fields Q^{jl} satisfying the bounds*

$$\|\nabla^k Q\|_{C^0} \le A(N\Xi)^k \left[\left(\frac{e_v^{1/2}}{e_R^{1/2} N} \right)^{-1/2} \frac{e_v^{1/2} e_R^{1/2}}{N} \right], \qquad k = 1, \dots, L \quad (889)$$

and for $k = 0, \dots, L-1$

$$\|\nabla^k (\partial_t + v_\epsilon \cdot \nabla) Q\|_{C^0} \le A(N\Xi)^k \left[\left(\frac{e_v^{1/2}}{e_R^{1/2} N} \right)^{-1/2} \Xi e_v^{1/2} e_R \right] \quad (890)$$

then we also have for $k = 0, \dots, L-1$

$$\|\nabla^k (\partial_t + v_1 \cdot \nabla) Q\|_{C^0} \le CA(N\Xi)^k \left[\left(\frac{e_v^{1/2}}{e_R^{1/2} N} \right)^{-1/2} \Xi e_v^{1/2} e_R \right]. \quad (891)$$

Proof. By writing

$$(\partial_t + v_1 \cdot \nabla) Q = (\partial_t + v_\epsilon \cdot \nabla) Q + [(v - v_\epsilon) + V] \cdot \nabla Q \quad (892)$$

and using (890) it suffices to prove the bound

$$\|\nabla^k [[(v - v_\epsilon) + V] \cdot \nabla Q]\|_{C^0} \le CA(N\Xi)^k \left[\left(\frac{e_v^{1/2}}{e_R^{1/2} N} \right)^{-1/2} \Xi e_v^{1/2} e_R \right] \quad (893)$$

for $k = 0, \dots, L-1$. For $k = 0$, using (470), Proposition (22.2), and the assumption (889) we have

$$\|[(v - v_\epsilon) + V] \cdot \nabla Q\|_{C^0} \le (\|v - v_\epsilon\|_{C^0} + \|V\|_{C^0}) \|\nabla Q\|_{C^0} \quad (894)$$

$$\le C \left(\frac{e_v^{1/2}}{N} + e_R^{1/2} \right) \left(A(N\Xi) \left(\frac{e_v^{1/2}}{e_R^{1/2} N} \right)^{-1/2} \frac{e_v^{1/2} e_R^{1/2}}{N} \right) \quad (895)$$

$$\le CA \left(\frac{e_v^{1/2}}{e_R^{1/2} N} \right)^{-1/2} (\Xi e_v^{1/2} e_R) \quad (896)$$

since $N \ge \left(\frac{e_v}{e_R} \right)^{1/2}$. Taking a spatial derivative will cost a factor $CN\Xi$ in the estimate. For example, if we estimate $\|\nabla(v - v_\epsilon)\|_{C^0}$ slightly suboptimally and we compare

$$\|(v - v_\epsilon)\|_{C^0} \le \frac{e_v^{1/2}}{N} \quad (897)$$

$$\|\nabla(v - v_\epsilon)\|_{C^0} \le \|\nabla v\|_{C^0} + \|\nabla v_\epsilon\|_{C^0} \le C\Xi e_v^{1/2} \quad (898)$$

we see that we lose a factor $CN\Xi$ by taking the derivative. For the whole product we have the bound

$$\|\nabla[[(v-v_\epsilon)+V]\cdot\nabla Q]\|_{C^0} \leq (\|\nabla v\|_{C^0}+\|\nabla v_\epsilon\|_{C^0}+s\|\nabla V\|_{C^0})\|\nabla Q\|_{C^0}$$
$$+ \|[(v - v_\epsilon) + V]\|_{C^0}\|\nabla^2 Q\|_{C^0}, \tag{899}$$

which is bounded by

$$\leq C\left(\Xi e_v^{1/2} + N\Xi e_R^{1/2}\right)\left(A\left(\frac{e_v^{1/2}}{e_R^{1/2}N}\right)^{-1/2}\Xi e_v^{1/2}e_R^{1/2}\right) \tag{900}$$

$$+ Ce_R^{1/2}\left(A(N\Xi)\left(\frac{e_v^{1/2}}{e_R^{1/2}N}\right)^{-1/2}\Xi e_v^{1/2}e_R^{1/2}\right) \tag{901}$$

$$\leq CA\left(\frac{e_v^{1/2}}{e_R^{1/2}N}\right)^{-1/2}(N\Xi)(\Xi e_v^{1/2}e_R) \tag{902}$$

using (889), Corollary 22.2, and the fact that $N \geq \left(\frac{e_v}{e_R}\right)^{1/2}$.

The assumption (889) and the bounds for v, v_ϵ and V also ensure that each higher spatial derivative costs at most $CN\Xi$ per derivative up to order $L-1$, which is what is required for the bound (891). □

The same proof also shows that it also suffices to verify estimates for the derivative $(\partial_t + v \cdot \nabla)Q$.

Proposition 24.2 (Material Derivative Criterion 2). *The result of Proposition (24.1) also holds with the condition (890) replaced by*

$$\|\nabla^k(\partial_t + v \cdot \nabla)Q\|_{C^0} \leq A(N\Xi)^k\left[B_\lambda^{1/2}\left(\frac{e_v^{1/2}}{e_R^{1/2}N}\right)^{-1/2}\Xi e_v^{1/2}e_R\right] \tag{903}$$

for all $k = 0, \ldots, L-1$.

We begin verifying the bounds (883)–(886) by estimating the terms that do not require solving the divergence equation.

25 Stress Terms Not Involving Solving the Divergence Equation

In this section, we estimate the terms in the new stress that do not involve solving the divergence equation. These terms include:

1. The Mollification Terms

$$Q_{M,v}^{jl} = (v^j - v_\epsilon^j)V^l + V^j(v^l - v_\epsilon^l) \tag{904}$$

$$Q_{M,R}^{jl} = R^{jl} - R_\epsilon^{jl} \tag{905}$$

2. The Stress Term

$$Q_S^{jl} = \sum_I \frac{(\nabla \times w_I)^j \bar{v}_I^l}{\lambda} + \sum_I \frac{v_I^j (\nabla \times \bar{w}_I)^l}{\lambda}$$
$$+ \sum_I \frac{(\nabla \times w_I)^j (\nabla \times \bar{w}_I)^l}{\lambda^2} \tag{906}$$

Throughout the estimates, we will assume that B_λ has been chosen to be some constant. We start with the Mollification terms.

25.1 The Mollification Term from the Velocity

The part of the new stress R_1 which arises from mollifying the velocity is given by

$$Q_{M,v}^{jl} = (v^j - v_\epsilon^j)V^l + V^j(v^l - v_\epsilon^l). \tag{907}$$

The C^0 estimate for $Q_{M,v}$ has already been verified by the choice of ϵ_v in Section (19). In fact, by verifying (465) we have that for the main term

$$Q_{v,(i)}^{jl} = \sum_I 2[(v - v_\epsilon)(e^{i\lambda\xi_I}v_I)]^{jl} \tag{908}$$

we have

$$\|Q_{v,(i)}^{jl}\|_{C^0} \le \frac{1}{50}\frac{e_v^{1/2}e_R^{1/2}}{N} \tag{909}$$

from our previous choice of the small constant c in the estimate (470), which we restate here as

$$\|(v^j - v_\epsilon^j)\|_{C^0} \le c\frac{e_v^{1/2}}{N}. \tag{910}$$

The other term for $Q_{M,v}$ is of the form

$$Q_{v,(ii)} = \sum_I 2e^{i\lambda\xi_I}\frac{[(v - v_\epsilon)\nabla \times w_I]^{jl}}{\lambda} \tag{911}$$

$$\|Q_{v,(ii)}\|_{C^0} \le C\left(\frac{e_v^{1/2}}{N}\right)\left(\frac{\Xi e_R^{1/2}}{B_\lambda N\Xi}\right) \tag{912}$$

which satisfies the goal

$$\|Q^{jl}_{v,(ii)}\|_{C^0} \le \frac{1}{50}\frac{e_v^{1/2}e_R^{1/2}}{N} \tag{913}$$

once the constant B_λ is chosen large enough.

We need to check the spatial derivatives of $Q_{M,v}$. By Definition (10.1), the estimates for spatial derivatives are allowed to lose a factor

$$|\nabla| \le \Xi' = CN\Xi$$

for each spatial derivative ∇^k, $k \le L$. This freedom allows us to separate the factors v and v_ϵ to bound

$$\nabla Q^{jl}_{M,v} = (\nabla vV)^{jl} - (\nabla v_\epsilon V)^{jl} + [(v-v_\epsilon)\nabla V]^{jl} \tag{914}$$

$$\|\nabla Q_{M,v}\|_{C^0} \le C(\|\nabla v\|_{C^0} + \|\nabla v_\epsilon\|_{C^0})e_R^{1/2} + \|v-v_\epsilon\|_{C^0}\|\nabla V\|_{C^0} \tag{915}$$

$$\le C\left(\Xi e_v^{1/2}e_R^{1/2} + \frac{e_v^{1/2}}{N}(B_\lambda N\Xi e_v^{1/2})\right) \tag{916}$$

$$\le CB_\lambda(\Xi e_v^{1/2})e_R^{1/2}. \tag{917}$$

By comparing

$$\|(v^j - v_\epsilon^j)\|_{C^0} \le c\frac{e_v^{1/2}}{N} \tag{918}$$

$$\|\nabla v\|_{C^0} + \|\nabla v_\epsilon\|_{C^0} \le 2\Xi e_v^{1/2} \tag{919}$$

we can see that the cost of separating v and v_ϵ in the estimate is exactly the factor $\Xi' = CN\Xi$ that we can afford. This principle will be useful many times in checking the inequalities.

By taking up to $L-1$ derivatives of (914), we see that for all $k = 1,\ldots,L$ we have the following.

Proposition 25.1. *For $k = 0,\ldots,L$, there exist constants C_k depending on B_λ such that*

$$\|\nabla^k Q_{M,v}\|_{C^0} \le C_k(B_\lambda N\Xi)^k\frac{e_v^{1/2}e_R^{1/2}}{N}. \tag{920}$$

This estimate is actually sufficient for the ideal case discussed in Section (13).

We now turn to estimating the material derivative.

Proposition 25.2. *Let $v_1 = v + V$. Then, for $k = 0,\ldots,L-1$, we have*

$$\|\nabla^k(\partial_t + v_1 \cdot \nabla)Q_{M,v}\|_{C^0} \le C\left(\frac{e_v^{1/2}}{e_R^{1/2}N}\right)^{-1/2}(N\Xi)^k\frac{e_v^{1/2}e_R^{1/2}}{N}. \tag{921}$$

To prove this proposition, it actually suffices to estimate $\frac{\bar{D}Q_{M,v}}{\partial t}$.

Proposition 25.3. *For $k = 0, \ldots, L-1$, we have*

$$\|\nabla^k(\partial_t + v_\epsilon \cdot \nabla)Q_{M,v}\|_{C^0} \leq C \left(\frac{e_v^{1/2}}{e_R^{1/2}N}\right)^{-1/2} (N\Xi)^k \frac{e_v^{1/2}e_R^{1/2}}{N} \tag{922}$$

by the criterion in Proposition (24.1). Now let us check the bounds (922).

Proof. Here we take

$$\frac{\bar{D}}{\partial t}[(v - v_\epsilon)V]^{jl} = \left(\frac{\bar{D}v}{\partial t}V\right)^{jl} - \left(\frac{\bar{D}v_\epsilon}{\partial t}V\right)^{jl} + [(v - v_\epsilon)\frac{\bar{D}V}{\partial t}]^{jl} \tag{923}$$

$$= A_{(I)} + A_{(II)} + A_{(III)}. \tag{924}$$

First, we compare

$$\|A_{(III)}\|_{C^0} \leq C\|v - v_\epsilon\|_{C^0}\|\frac{\bar{D}V}{\partial t}\|_{C^0} \tag{925}$$

$$\leq C\frac{e_v^{1/2}}{N}\left(\left(\frac{e_v^{1/2}}{e_R^{1/2}N}\right)^{-1/2}\Xi e_v^{1/2}e_R^{1/2}\right) \tag{926}$$

$$\leq C\left(\frac{e_v^{1/2}}{e_R^{1/2}N}\right)^{-1/2}\frac{\Xi e_v e_R^{1/2}}{N} \tag{927}$$

to our goal of

$$\|A_{(III)}\|_{C^0} \trianglelefteq \Xi'(e_v')^{1/2}e_R' \tag{928}$$

$$= C(N\Xi)e_R^{1/2}\left(\left(\frac{e_v^{1/2}}{e_R^{1/2}N}\right)^{-1/2}\frac{e_v^{1/2}e_R^{1/2}}{N}\right) \tag{929}$$

$$= C\left(\frac{e_v^{1/2}}{e_R^{1/2}N}\right)^{-1/2}\Xi e_v^{1/2}e_R \tag{930}$$

to see that $\|A_{(III)}\|_{C^0}$ has a correct bound since $N \geq \left(\frac{e_v}{e_R}\right)^{1/2}$. As we discussed in verifying (921), the estimates for

$$\|\nabla^k A_{(III)}\|_{C^0} \leq C_k(N\Xi)^k\left(\frac{e_v^{1/2}}{e_R^{1/2}N}\right)^{-1/2}\Xi e_v^{1/2}e_R$$

also hold for $k = 1, \ldots, L-1$, since each spatial derivative costs at most $N\Xi$, and because at most L derivatives fall on the given v.

Next we check how the bound

$$\|A_{(II)}\|_{C^0} \leq \|\frac{\bar{D}v_\epsilon}{\partial t}\|_{C^0}\|V\|_{C^0} \tag{931}$$

$$\leq (\Xi e_v)e_R^{1/2} \tag{932}$$

compares to the goal

$$\|A_{(II)}\|_{C^0} \trianglelefteq C \left(\frac{e_v^{1/2}}{e_R^{1/2}N}\right)^{-1/2} \Xi e_v^{1/2}e_R \tag{933}$$

from (930).

Checking this bound reduces to verifying that

$$\left(\frac{e_v}{e_R}\right)^{3/4} \leq N^{1/2} \tag{934}$$

which follows from the hypothesis

$$\left(\frac{e_v}{e_R}\right)^{3/2} \leq N \tag{935}$$

in Line (252) of the Main Lemma.

Now that the estimate for $A_{(II)}$ has been satisfied at the level of C^0, it worsens by a factor of $N\Xi$ for each spatial derivative, the largest cost arising when the derivative hits the oscillatory factor in the correction V. Therefore we have the bound

$$\|\nabla^k A_{(II)}\|_{C^0} \leq C(N\Xi)^k(\Xi e_v)e_R^{1/2}. \tag{936}$$

The goal

$$\|\nabla^k A_{(II)}\|_{C^0} \trianglelefteq C(N\Xi)^k \left(\frac{e_v^{1/2}}{e_R^{1/2}N}\right)^{-1/2} \Xi e_v^{1/2}e_R \qquad k = 1,\ldots,L \tag{937}$$

permits a loss of $CN\Xi$ per derivative, so that the bound (936) is acceptable.

We proceed similarly to estimate

$$A_{(I)} = \frac{\bar{D}v}{\partial t}V^l \tag{938}$$

by writing

$$\frac{\bar{D}v^l}{\partial t} = (\partial_t + v \cdot \nabla)v^l + (v_\epsilon - v) \cdot \nabla v^l \tag{939}$$

$$= \partial^l p + \partial_j R^{jl} + (v_\epsilon - v) \cdot \nabla v^l \tag{940}$$

159

giving

$$\|A_{(I)}\|_{C^0} \le C\|\frac{\bar{D}v}{\partial t}\|_{C^0} e_R^{1/2} \tag{941}$$

$$\le C(\Xi e_v + \left(\frac{e_v^{1/2}}{N}\right)(\Xi e_v^{1/2}))e_R^{1/2} \tag{942}$$

$$\le C\Xi e_v e_R^{1/2} \tag{943}$$

which we have just verified is acceptable for the term $A_{(II)}$. We can now take spatial derivatives of this term up to order $L - 1$, and each spatial derivative costs at most $CN\Xi$ per derivative; for example, comparing

$$\|v - v_\epsilon\|_{C^0} \le C\frac{e_v^{1/2}}{N} \tag{944}$$

$$\|\nabla(v - v_\epsilon)\|_{C^0} \le C\|\nabla v\|_{C^0} + \|\nabla v_\epsilon\|_{C^0} \le 2\Xi e_v^{1/2} \tag{945}$$

costs a factor $|\nabla| \le CN\Xi$. $\square$

Wastefulness in the Estimate To prove the ideal case outlined in Conjecture (10.1), the bound (922) must be replaced by the more stringent goal

$$\left\|\frac{\bar{D}}{\partial t}[(v - v_\epsilon)V]^{jl}\right\|_{C^0} \trianglelefteq C\Xi e_v^{1/2} e_R. \tag{946}$$

In order to verify inequality (946), it is necessary to establish a bound

$$\left\|\frac{\bar{D}}{\partial t}[v - v_\epsilon]\right\|_{C^0} \le C\Xi e_v^{1/2} e_R^{1/2} \tag{947}$$

which is stronger than the bound of Ξe_v that can be proven for $\|\frac{\bar{D}v}{\partial t}\|_{C^0}$ and $\|\frac{\bar{D}v_\epsilon}{\partial t}\|_{C^0}$ individually.

Inequality (947) can be proven through a straightforward but long commutator argument similar to those in Sections (16) and (18.6.2). Here we only note that this commutator estimate requires control over $\nabla[(\partial_t + v \cdot \nabla)v]$, and suggests that it may be important to work with Frequency Energy levels of order $L \ge 2$.

Here we have taken the slightly easier route of establishing the weaker bound (933), which has required imposing the lower bound (935) on N. The previous bounds are actually the only place in the argument in which the requirement (935) is used, and Lemma (10.1) can be proven with (935) replaced by

$$N \ge \left(\frac{e_v}{e_R}\right).$$

We also remark that in the ideal case construction, inequality (935) is satisfied during the iteration of Section (13), but in that case the power $3/2$ cannot be replaced by any larger value. During the iteration of Section (11), the value of N is essentially $\left(\frac{e_v}{e_R}\right)^{5/2}$.

25.2   The Mollification Term from the Stress

We now verify the necessary estimates for the term

$$Q_{M,R}^{jl} = R^{jl} - R_\epsilon^{jl}. \tag{948}$$

We have already verified in Section (18.3) that

$$\|Q_{M,R}^{jl}\|_{C^0} \le \frac{e_v^{1/2} e_R^{1/2}}{100N}, \tag{949}$$

and regarding the spatial derivatives, we have for $k = 1, \ldots, L$

$$\|\nabla^k Q_{M,R}\|_{C^0} \le \|\nabla^k R^{jl}\|_{C^0} + \|\nabla^k R_\epsilon^{jl}\|_{C^0} \tag{950}$$

$$\le C(\Xi^k e_R + \Xi^k e_R) \tag{951}$$

$$\le C\Xi^k e_R \tag{952}$$

from Proposition (18.3). Here, no powers of $N^{1/L}$ arise since we have taken no more than L derivatives. Comparing to our goal of

$$\|\nabla^k Q_{M,R}\|_{C^0} \trianglelefteq (\Xi')^k e_R' \tag{953}$$

$$= \left(\frac{e_v^{1/2}}{e_R^{1/2} N} \right)^{-1/2} (N\Xi)^k \frac{e_v^{1/2} e_R^{1/2}}{N} \tag{954}$$

we see that this goal has been achieved by a large margin.

We now check the bounds for the material derivative, which are of the following form.

Proposition 25.4. *For* $v_1 = v + V$, $k = 0, \ldots, L - 1$ *we have*

$$\|\nabla^k(\partial_t + v_1 \cdot \nabla)Q_{M,R}\|_{C^0} \le C(N\Xi)^k \Xi e_v^{1/2} e_R. \tag{955}$$

For our purposes, it is enough to recall the following from (244) and Proposition (18.4).

Proposition 25.5. *For* $0 \le k \le L - 1$,

$$\|\nabla^k(\partial_t + v \cdot \nabla)R\|_{C^0} \le C\Xi^{k+1} e_v^{1/2} e_R \tag{956}$$

$$\|\nabla^k(\partial_t + v_\epsilon \cdot \nabla)R_\epsilon\|_{C^0} \le C\Xi^{k+1} e_v^{1/2} e_R. \tag{957}$$

according to the criteria stated in Propositions (24.1) and (24.2).
We need to check, at least for $k = 0$, that we have achieved the goal

$$\|(\partial_t + v_1 \cdot \nabla)Q_{M,R}\|_{C^0} \trianglelefteq C\left(\Xi'(e_v')^{1/2} e_R' \right) \tag{958}$$

$$\trianglelefteq C \left(\frac{e_v^{1/2}}{e_R^{1/2} N} \right)^{-1/2} \left(N\Xi e_R^{1/2} \frac{e_v^{1/2} e_R^{1/2}}{N} \right) \tag{959}$$

$$= C \left(\frac{e_v^{1/2}}{e_R^{1/2} N} \right)^{-1/2} \Xi e_v^{1/2} e_R \tag{960}$$

In fact, this goal has been achieved even for the ideal case without the factor $\left(\frac{e_v^{1/2}}{e_R^{1/2}N}\right)^{-1/2}$. For spatial derivatives the estimates (955) show that each spatial derivative costs at most $N\Xi$ as desired, the largest losses coming from $V \cdot \nabla(R - R_\epsilon)$.

25.3   Estimates for the Stress Term

We now verify the estimates (883)–(886) for the Stress term, which turns out to be the most straightforward term to estimate.

$$
Q_S^{jl} = \sum_I \frac{(\nabla \times w_I)^j \bar{v}_I^l}{\lambda} + \sum_I \frac{v_I^j (\nabla \times \bar{w}_I)^l}{\lambda}
$$
$$
+ \sum_I \frac{(\nabla \times w_I)^j (\nabla \times \bar{w}_I)^l}{\lambda^2}
\tag{961}
$$

First, we have the bound

$$
\|Q_S\|_{C^0} \le C \frac{\Xi e_R^{1/2} \cdot e_R^{1/2}}{B_\lambda N \Xi} = C \frac{e_R}{B_\lambda N}
\tag{962}
$$

which implies that the goal for the ideal case

$$
\|Q_S\|_{C^0} \le \frac{e_v^{1/2} e_R^{1/2}}{500 N}
\tag{963}
$$

is satisfied for B_λ sufficiently large. From (961), we also have the estimates

$$
\|\nabla^k Q_S\|_{C^0} \le C_k (N\Xi)^k \frac{e_R}{N}
\tag{964}
$$

for $k = 1, \ldots, L$ (in fact, the bounds from Section (21) are much better), which implies the desired cost of $|\nabla| \le C N\Xi$ demanded by the goal (884).

Next, we have an estimate

$$
\|(\partial_t + v_\epsilon \cdot \nabla)Q_S\|_{C^0} \le C \left(\frac{e_v^{1/2}}{e_R^{1/2}N}\right)^{-1/2} \Xi e_v^{1/2} \frac{e_R}{N}
\tag{965}
$$

coming from the estimates in Section (21), which already implies the bound

$$
\|(\partial_t + v_\epsilon \cdot \nabla)Q_S\|_{C^0} \le C \Xi e_v^{1/2} e_R
\tag{966}
$$

which is desired for the ideal case as in (946). In particular, the bound (965) is enough to verify the required bound (885). It also follows from the estimates of Section (21) that the coarse scale material derivative $\frac{\bar{D}}{\partial t}Q_S$ can be differentiated in space arbitrarily many times at a cost no larger than $|\nabla| \le C N\Xi$ per derivative. By Proposition (24.1), it follows that all of the required estimates (883)–(886) are satisfied by Q_S with room to spare.

26 Terms Involving the Divergence Equation

In this section, we will estimate the terms in the stress which involve solving a divergence equation of the form

$$\partial_j Q^{jl} = U^l = e^{i\lambda\xi} u^l. \tag{967}$$

These terms include:

1. **The High-Low Interaction Term**

$$\partial_j Q_L^{jl} = \sum_I e^{i\lambda\xi_I} (\tilde{v}_I^j \partial_j v_\epsilon^l) \tag{968}$$

$$= \sum_I e^{i\lambda\xi_I} u_L^l \tag{969}$$

2. **The Main High-High Terms**

$$\partial_j Q_H^{jl} = -\sum_{J \neq \bar{I}} \lambda e^{i\lambda(\xi_I + \xi_J)} [v_I \times (|\nabla\xi_J| - 1) v_J^l + v_J \times (|\nabla\xi_I| - 1) v_I^l] \tag{970}$$

$$= \sum_{J \neq \bar{I}} u_{H,(IJ)}^l \tag{971}$$

3. **The Remainder of the High-High Terms**

$$\partial_j Q_{H'}^{jl}$$
$$= -e^{i\lambda(\xi_I + \xi_J)} \left((\nabla \times w_I) \times [(i\nabla\xi_J) \times \tilde{v}_J] + (\nabla \times w_J) \times [(i\nabla\xi_I) \times \tilde{v}_I] \right)$$
$$- e^{i\lambda(\xi_I + \xi_J)} \left(v_I \times [(i\nabla\xi_J) \times (\nabla \times w_J)] + v_J \times [(i\nabla\xi_I) \times (\nabla \times w_I)] \right)$$
$$- e^{i\lambda(\xi_I + \xi_J)} \left(\tilde{v}_I \times (\nabla \times \tilde{v}_J) + \tilde{v}_J \times (\nabla \times \tilde{v}_I) \right)$$
$$\tag{972}$$

4. **The Transport Term**

$$\partial_j Q_T^{jl} = \sum_I e^{i\lambda\xi_I} [(\partial_t + v_\epsilon^j \partial_j) \tilde{v}_I^l]$$
$$= \sum_I e^{i\lambda\xi_I} u_{T,(I)}^l \tag{973}$$

For each of these factors, we will use the parametrix expansion for the divergence equation, whose first term has the form

$$\partial_j Q^{jl} = e^{i\lambda\xi_I} u_1^l \tag{974}$$

$$Q^{jl} = \tilde{Q}_{(1)}^{jl} + Q_{(1)}^{jl} \tag{975}$$

$$\tilde{Q}_{(1)}^{jl} = e^{i\lambda\xi_I} \frac{q^{jl}(\nabla\xi)[u^l]}{\lambda} \tag{976}$$

$$\partial_j Q_{(1)}^{jl} = -e^{i\lambda\xi_I} \frac{1}{\lambda} \partial_j [q^{jl}(\nabla\xi)[u^l]] \tag{977}$$

and whose D'th term, in general, has the form

$$Q^{jl} = \tilde{Q}^{jl}_{(D)} + Q^{jl}_{(D)} \tag{978}$$

$$\tilde{Q}^{jl}_{(D)} = \sum_{(k)=1}^{D} e^{i\lambda\xi_I} \frac{q^{jl}_{(k)}}{\lambda^k} \tag{979}$$

where each $q_{(k)}$ solves a linear equation

$$i\partial_j\xi q^{jl}_{(1)} = u^l \tag{980}$$

$$i\partial_j\xi q^{jl}_{(k)} = u^l_{(k)} \tag{981}$$

$$u^l_{(k+1)} = -\partial_j q^{jl}_{(k)} \qquad 1 \le k \le D \tag{982}$$

using the linear map

$$q^{jl} = q^{jl}(\nabla\xi)[u] \tag{983}$$

defined in Line (38).

The error of the expansion is then eliminated by solving the divergence equation

$$\partial_j Q^{jl}_{(D)} = e^{i\lambda\xi} \frac{u^l_{(D+1)}}{\lambda^D}. \tag{984}$$

To eliminate the error, we need to solve the divergence equation, Equation (984), in such a way that we can have bounds for both $\|\nabla^k Q\|_{C^0}$ and $\|\nabla^k \frac{\tilde{D}Q}{\partial t}\|_{C^0}$. This task is accomplished in Section (27).

Let us first study the bounds which are obeyed for the parametrices of these oscillatory terms.

26.1 Expanding the Parametrix

For concreteness, consider the High-Low term

$$\partial_j Q^{jl}_L = \sum_I e^{i\lambda\xi_I} u^l_{L,I,(1)} \tag{985}$$

$$u^l_{L,I,(1)} = \tilde{v}^j_I \partial_j v^l_\epsilon. \tag{986}$$

Observe that

$$\|u_{L,I,(1)}\|_{C^0} \le \Xi e^{1/2}_v e^{1/2}_R \tag{987}$$

$$\|\nabla u_{L,I,(1)}\|_{C^0} \le (\|\nabla\tilde{v}_I\|_{C^0}\|\nabla v_\epsilon\|_{C^0} + \|\tilde{v}_I\|_{C^0}\|\nabla^2 v_\epsilon\|_{C^0}) \tag{988}$$

$$\le ((\Xi e^{1/2}_R)(\Xi e^{1/2}_v) + (e^{1/2}_R)(\Xi^2 e^{1/2}_v)) \tag{989}$$

$$\le C\Xi^2 e^{1/2}_v e^{1/2}_R \tag{990}$$

$$\|\nabla^k u_{L,I,(1)}\|_{C^0} \le \|\nabla^k \tilde{v}_I\|_{C^0}\|\nabla v_\epsilon\|_{C^0} + \ldots + \|\tilde{v}_I\|_{C^0}\|\nabla^{k+1} v_\epsilon\|_{C^0} \tag{991}$$

$$\le C(N^{1/L}\Xi)^k(\Xi e^{1/2}_v e^{1/2}_R) \tag{992}$$

from Section (21) and (474).

The first term in the parametrix has the form

$$\tilde{Q}_{L,I,(1)}^{jl} = e^{i\lambda\xi_I} \frac{q_{L,I,(1)}^{jl}}{\lambda} \tag{993}$$

$$q_{L,I,(1)}^{jl} = q^{jl}(\nabla\xi_I)[u_{L,I,(1)}^l]. \tag{994}$$

In size, the absolute value of the first term is

$$\|\tilde{Q}_{L,I,(1)}^{jl}\|_{C^0} \le C\frac{\|u_{L,I,(1)}\|_{C^0}}{\lambda} \tag{995}$$

$$\le C\frac{\Xi e_v^{1/2} e_R^{1/2}}{B_\lambda N\Xi} \tag{996}$$

$$\|\tilde{Q}_{L,I,(1)}^{jl}\|_{C^0} \le CB_\lambda^{-1}\frac{e_v^{1/2} e_R^{1/2}}{N} \tag{997}$$

which is actually good enough for the ideal theorem. Note that this is a gain of $\frac{1}{\lambda}$ compared to (987).

Having bounded the first term in the expansion, we now show that the later terms in the parametrix obey stronger bounds. Observe that the error term for the first expansion of the parametrix has the form

$$e^{i\lambda\xi_I} u_{L,I,(2)}^l = -e^{i\lambda\xi_I} \frac{\partial_j q_{L,I,(1)}^{jl}}{\lambda}. \tag{998}$$

Applying the chain rule to the function (994) defined in Line (38), we see that (998) can be bounded by

$$\frac{1}{\lambda}\|\nabla q_{L,I,(1)}^{jl}\|_{C^0} \le \frac{1}{\lambda}(\|\nabla^2\xi_I\|_{C^0}\|u_{L,I,(1)}\|_{C^0} + \|\nabla u_{L,I,(1)}\|_{C^0}) \tag{999}$$

$$\le \frac{C}{B_\lambda N\Xi}(\Xi(\Xi e_v^{1/2} e_R^{1/2}) + \Xi^2 e_v^{1/2} e_R^{1/2}) \tag{1000}$$

$$\le \frac{C}{B_\lambda N\Xi}(N^{1/L}\Xi^2)e_v^{1/2} e_R^{1/2}. \tag{1001}$$

Compared to $u_{L,I,(1)}$, the derivative ∂_j costs at most

$$|\partial_j| \le C\Xi N^{1/L}$$

whereas we gain a factor of

$$\lambda^{-1} = B_\lambda^{-1}\Xi^{-1}N^{-1}$$

giving us a gain of a factor

$$\frac{|\partial_j|}{\lambda^{-1}} \le \frac{C}{B_\lambda N^{(1-1/L)}} \tag{1002}$$

compared to the estimate for $u_{L,I,(1)}$. More precisely,

$$\lambda^{-1}\|u_{L,I,(2)}\|_{C^0} \leq CB_\lambda^{-1}\Xi^{-1}N^{-1}(\|\nabla^2\xi_I\|_{C^0}\|u_{L,I}\|_{C^0}+\|\nabla u_{L,I,(1)}\|_{C^0}) \tag{1003}$$

$$\leq CB_\lambda^{-1}N^{-(1-1/L)}\left(\Xi e_v^{1/2}e_R^{1/2}\right). \tag{1004}$$

This bound on the cost of a derivative continues to hold for higher derivatives, so that each iteration of the parametrix gains a smallness factor

$$\frac{|\nabla|}{\lambda} \leq C\frac{N^{1/L}\Xi}{B_\lambda N\Xi} = CB_\lambda^{-1}N^{-(1-1/L)}. \tag{1005}$$

For example,

$$\frac{1}{\lambda}\|\nabla^k q_{L,I,(1)}^{jl}\|_{C^0}$$

$$\leq \frac{C}{B_\lambda N\Xi}(\|\nabla^{k+1}\xi_I\|_{C^0}\|u_{L,I,(1)}\|_{C^0}+\ldots+\|\nabla^k\nabla u_{L,I,(1)}\|_{C^0}) \tag{1006}$$

$$\leq \frac{C}{B_\lambda N\Xi}(N^{1/L}\Xi)^k(\Xi e_v^{1/2}e_R^{1/2}). \tag{1007}$$

After iterating the parametrix D times, we have

$$\frac{1}{\lambda^D}\|u_{L,I,(D+1)}\|_{C^0} \leq C_D B_\lambda^{-D}N^{-D(1-1/L)}\left(\Xi e_v^{1/2}e_R^{1/2}\right). \tag{1008}$$

Thus, the estimate on the error improves by a factor $N^{-D(1-1/L)}$ after D iterations of the parametrix.

Because we have $N \geq \Xi^\eta$ for some $\eta > 0$, the gain of $N^{-D(1-1/L)}$ will eventually overtake the factor of Ξ, and we will be able to gain the factor $\frac{1}{B_\lambda N\Xi}$ that solving the elliptic equation is intended to gain. That is,

$$\frac{1}{\lambda^D}\|u_{L,I,(D+1)}\|_{C^0} \leq C_D B_\lambda^{-D}\frac{e_v^{1/2}e_R^{1/2}}{N} \tag{1009}$$

once D is large enough depending on the given η and L.

If we were only concerned about the C^0 norms and spatial derivatives, we could apply the operator $\mathcal{R}$ from Section (6) to obtain a solution to (984) of size

$$Q_{L,I,(D)} = -\frac{1}{\lambda^D}\mathcal{R}[e^{i\lambda\xi_I}u_{L,I,(D+1)}] \tag{1010}$$

$$\|Q_{L,I,(D)}\|_{C^0} \leq C\frac{\|u_{L,I,(D+1)}\|_{C^0}}{\lambda^D} \tag{1011}$$

$$\leq C_D B_\lambda^{-D}\frac{e_v^{1/2}e_R^{1/2}}{N} \tag{1012}$$

which can be bounded by

$$\|Q_{L,I,(D)}\|_{C^0} \le \frac{1}{1000}\frac{e_v^{1/2}e_R^{1/2}}{N} \tag{1013}$$

once B_λ is chosen large enough. Furthermore its derivatives would grow by a factor $(N\Xi)^k$ from

$$\|\nabla^k Q_{L,I,(D)}\|_{C^0} \le C_k \|\nabla^k \left[e^{i\lambda\xi_I}u_{L,I,(D)}\right]\|_{C^0} \tag{1014}$$

$$\le C_k \frac{1}{1000}(N\Xi)^k\frac{e_v^{1/2}e_R^{1/2}}{N}. \tag{1015}$$

However, we must also establish estimates for $\frac{\bar{D}Q_{L,I,(D)}}{\partial t}$, and for this purpose we have constructed a different operator to solve Equation (984).

Rather than the estimates for (1014), the solution which we construct in Section (27) to solve the equation

$$\partial_j Q_{L,I,(D)}^{jl} = U^l = \frac{-1}{\lambda^D}e^{i\lambda\xi_I}u_{L,I,(D+1)} \tag{1016}$$

is bounded by

$$\|Q_{L,I,(D)}\|_{C^0} \le C\|U\|_{C^0} + \tau\|\frac{\bar{D}U}{\partial t}\|_{C^0} \tag{1017}$$

$$\le \lambda^{-D}(\|u_{L,I,(D+1)}\|_{C^0} + \tau\|\frac{\bar{D}u_{L,I,(D+1)}}{\partial t}\|_{C^0}) \tag{1018}$$

and its support in time may grow to

$$\text{supp } Q_{L,I,(D)} \subseteq \{|t - t(I)| \le \frac{3\tau}{2}\} \times \mathbb{T}^3 \tag{1019}$$

when the support of U is contained in $|t - t(I)| \le \tau$. Even if the support does grow as in (1019), at each time t there will still be a bounded number of Stress terms $Q_{L,I,(D)}$ that are nonzero at that time.

Furthermore, this solution has the property that if the bounds on

$$\|\nabla^k U\|_{C^0} \le A(B_\lambda N\Xi)^k$$

$$\tau\|\nabla^k\frac{\bar{D}U}{\partial t}\|_{C^0} \le A(B_\lambda N\Xi)^k$$

are satisfied, then $Q_{L,I,(D)}$ obeys similar bounds

$$\|\nabla^k Q_{L,I,(D)}\|_{C^0} \le CA(B_\lambda N\Xi)^k \tag{1020}$$

$$\|\nabla^k\frac{\bar{D}Q_{L,I,(D)}}{\partial t}\|_{C^0} \le CA\tau^{-1}(B_\lambda N\Xi)^k. \tag{1021}$$

Namely, a spatial derivative of $Q_{L,I,(D)}$ costs $(B_\lambda N\Xi)$ and a material derivative of $Q_{L,I,(D)}$ costs τ^{-1} as long as these costs hold true for the data.

All of the examples of

$$U^l = e^{i\lambda\xi} u^l$$

that we will encounter share the feature that each material derivative costs τ^{-1}. Thus, even at the level of the material derivative, we will be able to conclude that each iteration of the parametrix gains a factor of

$$\frac{|\partial_j|}{\lambda} \leq N^{-(1-1/L)}.$$

Let us show now that this operator will be suitable for the conclusion of the proof.

26.2   Applying the Parametrix

First consider the Transport term, since it is the largest

$$\partial_j Q_{T,I}^{jl} = \sum_I e^{i\lambda\xi_I} u_{T,(I),1}^l \tag{1022}$$

$$u_{T,I,(1)}^l = (\partial_t + v_\epsilon^j \partial_j) \tilde{v}_I^l. \tag{1023}$$

Its parametrix expansion up to order D is equal to

$$Q_{T,I}^{jl} = \tilde{Q}_{T,I,(D)}^{jl} + Q_{T,I,(D)}^{jl} \tag{1024}$$

$$\tilde{Q}_{T,I,(D)}^{jl} = \sum_{k=1}^D e^{i\lambda\xi_I} \frac{q_{(k)}^{jl}}{\lambda^k} \tag{1025}$$

$$\partial_j Q_{T,I,(D)}^{jl} = \frac{-1}{\lambda^D} e^{i\lambda\xi_I} u_{T,I,(D+1)}^l. \tag{1026}$$

From Section (21), we have seen that each material derivative costs essentially τ^{-1}, so we know that

$$\|u_{T,I,(1)}^l\|_{C^0} + \tau \left\| \frac{\bar{D} u_{T,I,(1)}^l}{\partial t} \right\|_{C^0} \leq C B_\lambda^{1/2} \left(\frac{e_v^{1/2}}{e_R^{1/2} N} \right)^{-1/2} \Xi e_v^{1/2} e_R^{1/2} \tag{1027}$$

$$\|\nabla^k u_{T,I,(1)}^l\|_{C^0} + \tau \left\| \nabla^k \frac{\bar{D} u_{T,I,(1)}^l}{\partial t} \right\|_{C^0}$$

$$\leq C B_\lambda^{1/2} (B_\lambda N \Xi)^k \left(\frac{e_v^{1/2}}{e_R^{1/2} N} \right)^{-1/2} \Xi e_v^{1/2} e_R^{1/2}. \tag{1028}$$

The C^0 estimate was already checked in Section (19) where τ was chosen.

Using the estimates for spatial derivatives in (21) to commute transport and spatial derivatives, we also have

$$\lambda^{-D}\left(\|u^l_{T,I,(D+1)}\|_{C^0} + \tau \left\|\frac{\bar{D}u^l_{T,I,(D+1)}}{\partial t}\right\|_{C^0}\right) \le$$

$$\le C_D B_\lambda^{-D} N^{-D(1-1/L)} B_\lambda^{1/2}\left(\frac{e_v^{1/2}}{e_R^{1/2}N}\right)^{-1/2} \Xi e_v^{1/2} e_R^{1/2} \qquad (1029)$$

$$\lambda^{-D}\left(\|\nabla^k u^l_{T,I,(D+1)}\|_{C^0} + \tau \left\|\nabla^k\frac{\bar{D}u^l_{T,I,(D+1)}}{\partial t}\right\|_{C^0}\right) \le$$

$$\le C_D B_\lambda^{-D} N^{-D(1-1/L)} B_\lambda^{1/2}(B_\lambda N\Xi)^k\left(\frac{e_v^{1/2}}{e_R^{1/2}N}\right)^{-1/2} \Xi e_v^{1/2} e_R^{1/2}.$$

$$(1030)$$

Using (1029), choose D large enough depending on the η in the lower bound $N \ge \Xi^\eta$ such that

$$N^{-D(1-1/L)} \le N^{-1}\Xi^{-1}. \qquad (1031)$$

Namely, we expand the parametrix until we have gained the factor $N^{-1}\Xi^{-1}$ that we expect to gain from solving the divergence equation.

With this choice, the bounds (1029) and (1030) read

$$\lambda^{-D}\left(\|u^l_{T,I,(D+1)}\|_{C^0} + \tau \left\|\frac{\bar{D}u^l_{T,I,(D+1)}}{\partial t}\right\|_{C^0}\right)$$

$$\le C_D B_\lambda^{-D} B_\lambda^{1/2}\left(\frac{e_v^{1/2}}{e_R^{1/2}N}\right)^{-1/2} \frac{e_v^{1/2}e_R^{1/2}}{B_\lambda N} \qquad (1032)$$

$$\lambda^{-D}\left(\|\nabla^k u^l_{T,I,(D+1)}\|_{C^0} + \tau \left\|\nabla^k\frac{\bar{D}u^l_{T,I,(D+1)}}{\partial t}\right\|_{C^0}\right)$$

$$\le C_D B_\lambda^{-D} B_\lambda^{1/2}(B_\lambda N\Xi)^k\left(\frac{e_v^{1/2}}{e_R^{1/2}N}\right)^{-1/2} \frac{e_v^{1/2}e_R^{1/2}}{B_\lambda N}. \qquad (1033)$$

Applying Theorem (27.1), we have a solution $Q^{jl}_{D,T}$ to

$$\partial_j Q^{jl}_{T,I,(D)} = \sum_I \frac{-1}{\lambda^D} e^{i\lambda\xi_I} u^l_{T,I,(D+1)} \qquad (1034)$$

supported in $|t - t(I)| \leq 3\tau/2$ and obeying the bounds

$$\|\nabla^k Q_{T,I,(D)}\|_{C^0} + \tau \|\nabla^k \frac{\bar{D}Q_{T,I,(D)}}{\partial t}\|_{C^0}$$

$$\leq C_D B_\lambda^{1/2} (B_\lambda N\Xi)^k \left(\frac{e_v^{1/2}}{e_R^{1/2} N} \right)^{-1/2} \frac{e_v^{1/2} e_R^{1/2}}{B_\lambda N}. \tag{1035}$$

The parametrix itself

$$\tilde{Q}_{T,I,(D)}^{jl} = \sum_{(k)=1}^{D} e^{i\lambda \xi_I} \frac{q_{(k)}^{jl}}{\lambda^k} \tag{1036}$$

gains a factor of λ^{-1} compared to the first $u_{T,I,(1)}^l$, so it satisfies the bounds

$$\|\tilde{Q}_{T,I,(D)}^{jl}\|_{C^0} = C_D B_\lambda^{1/2} (B_\lambda N\Xi)^k \left(\frac{e_v^{1/2}}{e_R^{1/2} N} \right)^{-1/2} \frac{e_v^{1/2} e_R^{1/2}}{B_\lambda N} \tag{1037}$$

where the powers of $(B_\lambda N\Xi)$ come from differentiating the oscillatory factor.

Let us check that these are the estimates which are required in Lemma (10.1).

We essentially checked the C^0 bound for

$$\|\tilde{Q}_{T,I,(1)}^{jl}\|_{C^0} \leq C B_\lambda^{-1/2} \left(\frac{e_v^{1/2}}{e_R^{1/2} N} \right)^{-1/2} \frac{e_v^{1/2} e_R^{1/2}}{N} \tag{1038}$$

in Section (19). This estimate clearly worsens by a factor $B_\lambda N\Xi$ upon taking spatial derivatives, and worsens by a factor

$$|\frac{\bar{D}}{\partial t}| \leq \tau^{-1} = b^{-1} \Xi e_v^{1/2} \tag{1039}$$

$$= B_\lambda^{1/2} N^{1/2} \Xi e_R^{1/4} e_v^{1/4} \tag{1040}$$

upon taking a material derivative. Our desired cost for a material derivative is

$$|\frac{\bar{D}}{\partial t}| \trianglelefteq \Xi'(e_v')^{1/2} \tag{1041}$$

$$= CN\Xi e_R^{1/2} \tag{1042}$$

and checking this inequality is equivalent to verifying that

$$\left(\frac{e_v}{e_R} \right)^{1/4} \leq N^{1/2} \tag{1043}$$

$$\Leftrightarrow \left(\frac{e_v}{e_R} \right)^{1/2} \leq N \tag{1044}$$

which has been guaranteed in Lemma (10.1).

Every other Stress term can be treated similarly and exhibits even better bounds except for the main term in the High-High term, which is

$$\partial_j Q_H^{jl} = \sum_{J \neq \bar{I}} \lambda e^{i\lambda(\xi_I + \xi_J)} u_{H,IJ}^l \tag{1045}$$

$$u_{H,IJ}^l = -[v_I \times (|\nabla \xi_J| - 1)v_J^l + v_J \times (|\nabla \xi_I| - 1)v_I^l]. \tag{1046}$$

The first term in the parametrix expansion for this term takes the form

$$\tilde{Q}_{H,IJ,(1)}^{jl} = e^{i\lambda(\xi_I + \xi_J)} q^{jl}(\nabla(\xi_I + \xi_J))[u_{H,IJ}]. \tag{1047}$$

For this term, we observed in Section (19) that

$$\|\tilde{Q}_{H,IJ,(1)}^{jl}\|_{C^0} \leq C e_R b \tag{1048}$$

$$\leq C B_\lambda^{-1/2} \left(\frac{e_v^{1/2}}{e_R^{1/2} N} \right)^{1/2} e_R \tag{1049}$$

$$\leq C B_\lambda^{-1/2} \left(\frac{e_v^{1/2}}{e_R^{1/2} N} \right)^{-1/2} \frac{c_v^{1/2} e_R^{1/2}}{N} \tag{1050}$$

thanks to our choice of τ.

For this term one must also check that the cost of a spatial derivative is bounded below $N\Xi$. To check this bound, we compare

$$\||\nabla \xi_J| - 1\|_{C^0} \leq C B_\lambda^{-1/2} \left(\frac{e_v^{1/2}}{e_R^{1/2} N} \right)^{1/2} \tag{1051}$$

$$\|\nabla(|\nabla \xi_J| - 1)\|_{C^0} \leq C \|\nabla^2 \xi_J\|_{C^0} \leq C\Xi \tag{1052}$$

to reveal that differentiating worsens the bounds by at most a factor

$$|\nabla| \leq C B_\lambda^{1/2} \left(\frac{e_R^{1/2} N}{e_v^{1/2}} \right)^{1/2} \Xi \tag{1053}$$

$$|\nabla| \leq C B_\lambda^{1/2} N^{1/2} \Xi \tag{1054}$$

so that each iteration of the parametrix gains at least

$$\frac{|\nabla|}{\lambda} \leq C \frac{B_\lambda^{1/2} N^{1/2} \Xi}{B_\lambda N \Xi} \tag{1055}$$

$$\leq C B_\lambda^{-1/2} N^{-1/2} \tag{1056}$$

which will also gain the factor of $N^{-1}\Xi^{-1}$ which we require once sufficiently many terms in the parametrix expansion have been taken. Likewise, we must check that each material derivative of this term costs at most τ^{-1}, which follows from comparing the following inequalities

$$\||\nabla \xi_J| - 1\|_{C^0} \leq Cb = CB_\lambda^{-1/2} \left(\frac{e_v^{1/2}}{e_R^{1/2} N} \right)^{1/2} \tag{1057}$$

$$\|\frac{\bar{D}}{\partial t}(|\nabla \xi_J| - 1)\|_{C^0} \leq C\|\frac{\bar{D}}{\partial t}\nabla \xi_J\|_{C^0} \leq C\Xi e_v^{1/2} \tag{1058}$$

$$\Rightarrow \left| \frac{\bar{D}}{\partial t} \right| \leq CB_\lambda^{1/2} \left(\frac{e_v^{1/2}}{e_R^{1/2} N} \right)^{-1/2} (\Xi e_v^{1/2}) \tag{1059}$$

$$\leq Cb^{-1}\Xi e_v^{1/2} \leq C\tau^{-1}. \tag{1060}$$

Thus, the High-High term can be treated in the same way as the Transport term, taking more terms in the parametrix expansion if necessary. In each case, the terms Q_T, Q_H and Q_L enjoy a bound of the same type as the bound on the first term of their parametrix expansion but with a worse constant. Among all the terms composing the new stress, the terms Q_T and Q_H obey the worst bound of

$$\|Q_T\|_{C^0} + \|Q_H\|_{C^0} \leq CB_\lambda^{-1/2} \left(\frac{e_v^{1/2}}{e_R^{1/2} N} \right)^{-1/2} \frac{e_v^{1/2} e_R^{1/2}}{N}. \tag{1061}$$

In particular, we have a bound

$$\|R_1\|_{C^0} \leq CB_\lambda^{-1/2} \left(\frac{e_v^{1/2}}{e_R^{1/2} N} \right)^{-1/2} \frac{e_v^{1/2} e_R^{1/2}}{N} \tag{1062}$$

for the entirety of the new stress.

At this point, we can finally choose the constant B_λ so that the goal

$$\|R_1\|_{C^0} \leq CB_\lambda^{-1/2} \left(\frac{e_v^{1/2}}{e_R^{1/2} N} \right)^{-1/2} \frac{e_v^{1/2} e_R^{1/2}}{N} \tag{1063}$$

$$\trianglelefteq \frac{1}{20} \left(\frac{e_v^{1/2}}{e_R^{1/2} N} \right)^{-1/2} \frac{e_v^{1/2} e_R^{1/2}}{N} \tag{1064}$$

is satisfied.

Once B_λ has been chosen for this purpose, we know that each spatial derivative of R_1 costs at most

$$|\nabla| \leq CN\Xi$$

which is exactly the target cost for a spatial derivative. Each material derivative costs

$$\left| \frac{\bar{D}}{\partial t} \right| \leq C\tau^{-1} = C \left(\frac{e_v^{1/2}}{e_R^{1/2} N} \right)^{-1/2} \Xi e_v^{1/2} \tag{1065}$$

which must be below the target cost of

$$\left|\frac{\bar{D}}{\partial t}\right| \lhd \Xi'(e'_v)^{1/2} \tag{1066}$$

$$\lhd CN\Xi e_R^{1/2} \tag{1067}$$

and this target cost has been satisfied as long as

$$\left(\frac{e_v}{e_R}\right)^{1/4} \leq N^{1/2}. \tag{1068}$$

This inequality follows from the condition

$$N \geq \left(\frac{e_v}{e_R}\right)$$

and is enough to conclude the proof of Lemma (10.1).

27 Transport-Elliptic Estimates

In order to eliminate the error term in the parametrix it is necessary to solve the underdetermined, elliptic equation

$$\partial_j Q^{jl} = U^l$$
$$Q^{jl} = Q^{lj}. \tag{1069}$$

For the proof of the Main Lemma, we need to have estimates for Q, spatial derivatives $\nabla^k Q$ of Q, and also the material derivative $\frac{\bar{D}Q}{\partial t}$ and its spatial derivatives.

First we recall from Lemma (6.2) that the divergence operator in Equation (1069) can be inverted by an operator $\mathcal{R}$ that has order -1.

Rather than commuting $\frac{\bar{D}}{\partial t}$ with the nonlocal operator $\mathcal{R}$, we find a solution to (1069) with good transport properties by solving (1069) via a transport equation obtained by commuting the divergence operator with the material derivative. In this way, we directly obtain estimates for $\frac{\bar{D}Q}{\partial t}$ and its derivatives.

Theorem 27.1. *Let U^l be a vector field such that*

$$\int U^l(t, x) dx = 0 \tag{1070}$$

for all t, and such that

$$\mathrm{supp}\, U \subseteq [t(I) - \tau, t(I) + \tau] \times \mathbb{T}^3 \tag{1071}$$

for some time $t(I) \in \mathbb{R}$ and some $\tau \leq \Xi^{-1} e_v^{-1/2}$.

173

Assume also that for

$$\Lambda = B_\lambda N\Xi$$

the velocity field U and its material derivative $\frac{\bar{D}U}{\partial t}$ obey the estimates

$$\|\nabla^k U\|_{C_t^0 L_x^4} \leq A\Lambda^k \qquad |k| = 0, \ldots, L$$

$$\|\nabla^k \frac{\bar{D}U}{\partial t}\|_{C_t^0 L_x^4} \leq A\tau^{-1}\Lambda^k \qquad k = 0, \ldots, L-1. \tag{1072}$$

Then there exists a solution Q to Equation (1069) depending linearly on U such that

1. for all t,

$$\int Q(t,x)dx = 0 \tag{1073}$$

2.

$$\text{supp } Q \subseteq [t(I) - 3\tau/2, t(I) + 3\tau/2] \times \mathbb{T}^3 \tag{1074}$$

3. for $k = 0, \ldots, L$

$$\|\nabla^{k+1}Q\|_{C_t^0 L_x^4} \leq CA\Lambda^k \tag{1075}$$

4. for $k = 0, \ldots, L-1$

$$\|\nabla^{k+1}\frac{\bar{D}Q}{\partial t}\|_{C_t^0 L_x^4} \leq CA\tau^{-1}\Lambda^k. \tag{1076}$$

The proof of Theorem (27.1) relies on a transport equation which we now derive. To set up the initial data, we set

$$Q^{jl}(t(I), x) = \mathcal{R}^{jl}[U(t(I), \cdot)](x). \tag{1077}$$

With this choice of data, Equation (1069) is satisfied at the initial time $t(I)$. The divergence equation will remain satisfied at future times if we ensure that

$$\frac{\bar{D}}{\partial t}\partial_j Q^{jl} = (\partial_t + v_\epsilon^i \partial_i)\partial_j Q^{jl} = \frac{\bar{D}U^l}{\partial t}. \tag{1078}$$

Commuting $\frac{\bar{D}}{\partial t}$ with the divergence operator leads to an underdetermined elliptic equation

$$\frac{\bar{D}}{\partial t}\partial_j Q^{jl} = \partial_j\left[\frac{\bar{D}}{\partial t}Q^{jl}\right] - \partial_j v_\epsilon^i \partial_i Q^{jl} = \frac{\bar{D}U^l}{\partial t}. \tag{1079}$$

We solve the above equation by inverting the divergence operator, which leads to an equation that we call the Transport-Elliptic equation.

174

Lemma 27.1. *Suppose that U^l is a smooth vector field of integral 0, then if Q^{jl} solves the transport equation*

$$\frac{\bar{D}Q^{jl}}{\partial t} = \mathcal{R}^{jl}[\partial_i v_\epsilon^b \partial_b Q^{ik} + \frac{\bar{D}U^k}{\partial t}] \tag{1080}$$

with initial data

$$Q^{jl}(t(I), x) = \mathcal{R}^{jl}[U(t(I), \cdot)](x) \tag{1081}$$

then Q also solves the equation

$$\partial_j Q^{jl} = U^l \tag{1082}$$

with

$$\int Q dx = 0 \tag{1083}$$

for all t.

Proof. At the initial time $t = t(I)$, we have

$$\partial_j Q^{jl}(t(I), x) = \partial_j \mathcal{R}^{jl}[U(t(I), \cdot)] \tag{1084}$$
$$- \mathcal{P}U^l(t(I), x) \tag{1085}$$

where $\mathcal{P}$ is the operator which projects to integral 0 vector fields. Because U has integral 0, we have

$$\mathcal{P}U^l(t, x) = U^l(t, x)$$

so it is clear that Equation (1082) is satisfied at time $t(I)$.

It now suffices to verify that (1079) is satisfied by the uniqueness of solutions to the initial value problem for transport equations. This calculation will rely crucially on the fact that v_ϵ is divergence free, which implies, for example, that the term

$$\partial_j v_\epsilon^b \partial_b Q^{jl} = \partial_j \partial_b [v_\epsilon^b Q^{jl}] \tag{1086}$$

has integral 0.

By taking the divergence of (1080), we compute

$$\partial_j [\frac{\bar{D}}{\partial t} Q^{jl}] = \mathcal{P}[\partial_j v_\epsilon^b \partial_b Q^{jl}] + \mathcal{P}\frac{\bar{D}U^l}{\partial t} \tag{1087}$$

$$= \mathcal{P}[\partial_j \partial_b [v_\epsilon^b Q^{jl}]] + \mathcal{P}\frac{\bar{D}U^l}{\partial t} \tag{1088}$$

$$= \partial_j \partial_b [v_\epsilon^b Q^{jl}] + \mathcal{P}\frac{\bar{D}U^l}{\partial t}. \tag{1089}$$

175

Since U^l has integral 0 and v_ϵ is divergence free, we have that

$$\int \frac{\bar{D}U}{\partial t} dx = \frac{d}{dt} \int U dx = 0 \tag{1090}$$

implying that $\mathcal{P}\frac{\bar{D}U^l}{\partial t} = \frac{\bar{D}U^l}{\partial t}$.

From (1089), we have

$$\partial_j [\frac{\bar{D}}{\partial t} Q^{jl}] = \partial_j [v_\epsilon^b \partial_b Q^{jl}] + \frac{\bar{D}U^l}{\partial t} \tag{1091}$$

which is Equation (1079) that we had to verify.

To see that $\int Q dx = 0$ for all t, first observe that

$$\int Q(t(I), x) dx = 0 \tag{1092}$$

by the property that the operator $\mathcal{R}$ maps to integral 0 tensors, and for the same reason,

$$\frac{d}{dt} \int Q(t(I), x) dx = \int \frac{\bar{D}Q}{\partial t} dx = 0 \tag{1093}$$

as well. $\qquad\qquad\qquad\qquad\qquad\qquad\qquad\qquad\qquad\qquad\qquad\square$

Our next goal for the section is to establish existence and a priori estimates for the solution to the PDE (1080).

27.1 Existence of Solutions for the Transport-Elliptic Equation

Global solutions to the linear equation

$$\frac{\bar{D}Q^{jl}}{\partial t} = \mathcal{R}^{jl}[\partial_i v_\epsilon^b \partial_b Q^{ik} + \frac{\bar{D}U^k}{\partial t}]$$
$$Q^{jl}(t(I), x) = \mathcal{R}^{jl}[U(t(I), \cdot)](x) \tag{1094}$$

can be easily constructed by the method of Picard iteration, once the appropriate spaces have been identified.

To begin the Picard iteration we define an operator T acting on symmetric $(2, 0)$-tensor fields which solves the transport equation

$$\frac{\bar{D}}{\partial t}[TQ]^{jl} = \mathcal{R}^{jl}[\partial_i v_\epsilon^b \partial_b Q^{ik} + \frac{\bar{D}U^k}{\partial t}]$$
$$TQ^{jl}(t(I), x) = \mathcal{R}^{jl}[U(t(I), \cdot)](x) \tag{1095}$$

so that a fixed point of T is a solution to the initial value problem (1094).

The operator T must be defined on an appropriately defined complete metric space, and the key to identifying this metric space is the following, a priori estimate

Proposition 27.1 (A Priori Estimate). *There are constants $C_1, C_2 \geq 0$ such that if Q solves (1094) and U satisfies the bounds (1072), then*

1.

$$\int Q(t, x) dx = 0 \tag{1096}$$

2.

$$\sum_{|a|=1} \int |\partial_a Q|^4(t, x) dx \leq C_1 A^4 e^{C_2 \tau^{-1}|t-t(I)|}. \tag{1097}$$

Proof. At time $t = t(I)$, the bound (1097) is a consequence of the bound

$$\|\nabla \mathcal{R}[U]\|_{L^4} \leq C \|U\|_{L^4}$$

so by Gronwall it suffices to prove that

$$|\sum_{|a|=1} \frac{d}{dt} \int |\partial_a Q|^4(t, x) dx| \leq C_2 \tau^{-1} (\sum_{|b|=1} \sum_{j,l=1}^{3} \int |\partial_b Q^{jl}|^4(t, x) dx + A^4). \tag{1098}$$

To establish this inequality, we let ∂_a be any derivative of order $|a| = 1$, and for any component $\partial_a Q^{jl}$ of $\partial_a Q$ we compute the time derivative

$$\frac{d}{dt} \int |\partial_a Q^{jl}|^4(t, x) dx = \int \frac{\bar{D}}{\partial t} |\partial_a Q^{jl}|^4 dx \tag{1099}$$

$$= \int 4(\partial_a Q^{jl})^3 [\frac{\bar{D}}{\partial t} \partial_a Q^{jl}] dx \tag{1100}$$

$$= \int 4(\partial_a Q^{jl})^3 [\partial_a (\frac{\bar{D}}{\partial t} Q^{jl}) - \partial_a v_\epsilon^i \partial_i Q^{jl}] dx. \tag{1101}$$

Since $\|\nabla v_\epsilon\|_{C^0} \leq \Xi e_v^{1/2} < \tau^{-1}$, the second of these terms can be bounded by

$$\left| 4 \int (\partial_a Q^{jl})^3 (\partial_a v_\epsilon^i \partial_i Q^{jl}) dx \right| \leq C \tau^{-1} \int |\partial_a Q^{jl}|^4(t, x) dx. \tag{1102}$$

For the first term in (1101), we use Equation (1094) to estimate

$$\left| \int (\partial_a Q^{jl})^3 [\partial_a (\frac{\bar{D}}{\partial t} Q^{jl})] dx \right|$$

$$= \left| \int (\partial_a Q^{jl})^3 \left(\partial_a \mathcal{R}^{jl} [\partial_i v_\epsilon^b \partial_b Q^{ik} + \frac{\bar{D} U^k}{\partial t}] \right) dx \right| \tag{1103}$$

$$\leq \int \left(|\partial_a Q^{jl}|^3 |\partial_a \mathcal{R}^{jl} [\partial_i v_\epsilon^b \partial_b Q^{ik}]| + |\partial_a Q|^3 |\partial_a \mathcal{R}^{jl} [\frac{\bar{D} U^k}{\partial t}]| \right) dx.$$

The first of these terms is similar to what we have already encountered, and can be estimated using Hölder's inequality.

$$\int |\partial_a Q^{jl}|^3 \left|\partial_a \mathcal{R}^{jl}[\partial_i v_\epsilon^b \partial_b Q^{ik}]\right| dx \le \|\nabla Q\|_{L^4}^3 \|\nabla \mathcal{R}^{jl}[\partial_i v_\epsilon^b \partial_b Q^{ik}]\|_{L^4} \quad (1104)$$

$$\le \|\nabla Q^{jl}\|_{L^4}^3 \|\partial_i v_\epsilon^b \partial_b Q^{ik}\|_{L^4} \quad (1105)$$

$$\le \Xi e_v^{1/2} \|\nabla Q\|_{L^4}^4 \quad (1106)$$

$$\le C\tau^{-1} \sum_{|b|=1} \sum_{j,l=1}^3 \int |\partial_b Q^{jl}|^4 (t,x) dx \quad (1107)$$

The latter term in (1103) can be estimated equivalently by applying Young's inequality with exponents $\frac{3}{4} + \frac{1}{4} = 1$ to the pointwise product.

$$|\partial_a Q^{jl}|^3 |\partial_a \mathcal{R}^{jl}[\frac{\bar{D}U}{\partial t}]| = (\tau^{-4/3}|\partial_a Q^{jl}|^3)(\tau^{3/4}|\partial_a \mathcal{R}^{jl}[\frac{\bar{D}U}{\partial t}]|) \quad (1108)$$

$$\le \frac{3}{4}\tau^{-1}|\partial_a Q^{jl}|^4 + \frac{1}{4}\tau^3 \left|\partial_a \mathcal{R}^{jl}[\frac{\bar{D}U}{\partial t}]\right|^4 \quad (1109)$$

This calculation gives the bounds

$$\int |\partial_a Q^{jl}|^3 |\partial_a \mathcal{R}^{jl}[\frac{\bar{D}U}{\partial t}]| dx \le C \int (\tau^{-1}|\partial_a Q^{jl}|^4 + \tau^3 \left|\partial_a \mathcal{R}^{jl}[\frac{\bar{D}U}{\partial t}]\right|^4) dx$$
$$(1110)$$

$$\le C\tau^{-1} \left(\int |\partial_a Q^{jl}|^4 dx + \int \tau^4 \left|\frac{\bar{D}U}{\partial t}\right|^4 dx\right) \quad (1111)$$

$$\le C\tau^{-1} \left(\int |\partial_a Q^{jl}|^4 dx + A^4\right) \quad (1112)$$

which is the estimate required. $\qquad\square$

The proof of the above estimate can also be used to establish that on the space

$$X \subseteq C_t^0 W_x^{1,4}(\mathbb{R} \times \mathbb{T}^3; \mathcal{S}) \subseteq C_{t,x}^0(\mathbb{R} \times \mathbb{T}^3; \mathcal{S})$$

defined by the conditions

$$X = \{Q^{jl} : \mathbb{R} \times \mathbb{T}^3 \to \mathcal{S} \mid \int Q(t,x)dx = 0 \;\forall\, t \in \mathbb{R},$$

$$Q^{jl}(t(I),x) = \mathcal{R}^{jl}[U(t(I),\cdot)](x)$$

$$\sum_{|a|=1} \sum_{i,j=1}^3 \int |\partial_a Q^{ij}|^4 (t,x) dx \le C_1 A^4 e^{C_2 \tau^{-1}|t-t(I)|}\}$$

the map T defined in (1095) maps X to itself when C_1 and C_2 are chosen appropriately. Furthermore, when the space X is endowed with the metric deriving from the norm

$$\|Q\|_X = \sup_{t \in \mathbb{R}} e^{-B\tau^{-1}|t-t(I)|} \left(\sum_{|a|=1} \sum_{j,l=1}^{3} \int |\partial_a Q^{jl}|^4(t,x)dx \right)^{1/4} \quad (1113)$$

then X is a complete metric space, and essentially the same proof as that of Proposition (27.1) also shows that T is a contraction on X whenever B is a sufficiently large constant. Therefore a unique, global solution to the initial value problem (1094) exists.

27.2 Spatial Derivative Estimates for the Solution to the Transport-Elliptic Equation

Following the methods of Section (27.1), we can also give bounds on the higher derivatives of solutions to Equation (1094).

Proposition 27.2. *The solution Q to Equation (1094) obeys the bounds*

$$\|\nabla^{k+1}Q\|_{L_+^4} \le C_1 \Lambda^k e^{C_2 \tau^{-1}|t-t(I)|} A \quad (1114)$$

for $k = 0, \ldots, L$.

Proof. At the time $t(I)$, the above inequality follows from the bounds (47) for $\mathcal{R}$ and the assumed bounds (1072) on U and its spatial derivatives.

To establish the estimates (1114), we follow the strategy of Section (27.1) and define the following weighted energy.

Definition 27.1.

$$E_M[Q](t) \equiv \sum_{K=1}^{M} \sum_{j,l=1}^{3} \sum_{|a|=K} \int \left| \frac{\partial_a Q^{jl}}{\Lambda^{K-1}} \right|^4 dx \quad (1115)$$

Since

$$E_M[Q](t(I)) \le C_1 A^4,$$

the exponential bound (1114) follows from Gronwall after we establish a differential inequality

$$\left| \frac{dE_M[Q]}{dt} \right| \le C_2 \tau^{-1}(E_M[Q] + A^4). \quad (1116)$$

To establish the differential inequality, we fix a spatial derivative ∂_a of order K and components jl of $\partial_a Q^{jl}$ and calculate

$$\frac{d}{dt} \int \left| \frac{\partial_a Q^{jl}}{\Lambda^{K-1}} \right|^4 dx = \int \frac{\bar{D}}{\partial t} \left| \frac{\partial_a Q^{jl}}{\Lambda^{K-1}} \right|^4 dx \quad (1117)$$

$$= 4 \int \left(\frac{\partial_a Q^{jl}}{\Lambda^{K-1}} \right)^3 \cdot \Lambda^{-(K-1)} [\frac{\bar{D}}{\partial t} \partial_a Q^{jl}]. \quad (1118)$$

179

Commutator Terms As in Section (27.1), we now commute $\frac{\bar{D}}{\partial t} = (\partial_t + v_\epsilon^b \partial_b)$ with $\partial_a = \partial_{a_1} \cdots \partial_{a_K}$ to see that

$$\frac{\bar{D}}{\partial t} \partial_a Q^{jl} = \partial_a [\frac{\bar{D} Q^{jl}}{\partial t}] - \sum_{|a_1|+|a_2|=K, |a_1| \geq 1} C^a_{a_1, a_2} \partial_{a_1} v_\epsilon^b \partial_{a_2} \partial_b Q^{jl}. \quad (1119)$$

The commutator terms are acceptable in (1118), because they give rise to terms bounded by

$$\int \left| \frac{\partial_a Q^{jl}}{\Lambda^{K-1}} \right|^3 \cdot (\Lambda^{-(K-1)} |\partial_{a_1} v_\epsilon^b \partial_{a_2} \partial_b Q^{jl}|) dx$$

$$\leq \int \left| \frac{\partial_a Q^{jl}}{\Lambda^{K-1}} \right|^3 \left| \frac{\partial_{a_1} v_\epsilon^b}{\Lambda^{|a_1|-1}} \right| \left| \frac{\partial_{a_2} \partial_b Q^{jl}}{\Lambda^{|a_2|}} \right|. \quad (1120)$$

Because $|a_1| \geq 1$, we have $|a_2| \leq K - 1$, and we can estimate the above by

$$\int \left| \frac{\partial_a Q^{jl}}{\Lambda^{K-1}} \right|^3 \cdot (\Lambda^{-(K-1)} |\partial_{a_1} v_\epsilon^b \partial_{a_2} \partial_b Q^{jl}|) dx \leq \left\| \frac{\partial_{a_1} v_\epsilon^b}{\Lambda^{|a_1|-1}} \right\|_{C^0} E_M[Q](t) \quad (1121)$$

$$\leq C \frac{\Xi^{|a_1|}}{\Lambda^{|a_1|-1}} e_v^{1/2} E_M[Q](t) \quad (1122)$$

$$\leq C \Xi e_v^{1/2} E_M[Q](t) \quad (1123)$$

$$\leq C \tau^{-1} E_M[Q](t). \quad (1124)$$

Concluding the Estimate for (1118) To conclude the proof of (1116), it remains to show that for all multi-indices of order $|a| = K - 1$, we have

$$\left| \int \left(\frac{\partial_a Q^{jl}}{\Lambda^{K-1}} \right)^3 \cdot \Lambda^{-(K-1)} \left[\partial_a \frac{\bar{D} Q^{jl}}{\partial t} \right] dx \right| \leq C \tau^{-1} (E_M[Q](t) + A^4). \quad (1125)$$

Using Equation (1094), the integral inside the absolute values can be written as a sum

$$\int \left(\frac{\partial_a Q^{jl}}{\Lambda^{K-1}} \right)^3 \cdot \Lambda^{-(K-1)} [\partial_a \frac{\bar{D} Q^{jl}}{\partial t}] dx$$

$$= \int \left(\frac{\partial_a Q^{jl}}{\Lambda^{K-1}} \right)^3 \cdot \Lambda^{-(K-1)} \left(\partial_a \mathcal{R}^{jl} [\partial_i v_\epsilon^b \partial_b Q^{ik}] \right) dx \quad (1126)$$

$$+ \int \left(\frac{\partial_a Q^{jl}}{\Lambda^{K-1}} \right)^3 \cdot \Lambda^{-(K-1)} \partial_a \mathcal{R}^{jl} [\frac{\bar{D} U}{\partial t}] dx \quad (1127)$$

$$= I + II. \quad (1128)$$

Let us first estimate the term I. By Hölder's inequality, I is bounded by

$$|I| \leq \|\frac{\partial_a Q^{jl}}{\Lambda^{K-1}}\|_{L_x^4}^3 \cdot \Lambda^{-(K-1)} \|\partial_a \mathcal{R}^{jl}[\partial_i v_\epsilon^b \partial_b Q^{ik}]\|_{L_x^4}. \quad (1129)$$

By the boundedness properties (47) of $\mathcal{R}$, we have that

$$\Lambda^{-(K-1)} \|\partial_a \mathcal{R}^{jl}[\partial_i v_\epsilon^b \partial_b Q^{ik}]\|_{L_x^4} \leq C\Lambda^{-(K-1)} \|\nabla^{K-1}[\partial_i v_\epsilon^b \partial_b Q^{ik}]\|_{L_x^4} \quad (1130)$$

$$\leq C\Lambda^{-(K-1)} \sum_{|a_1|+|a_2|=K-1} \|\partial_{a_1} \nabla v_\epsilon^b\|_{C^0} \|\partial_{a_2} \partial_b Q\|_{L^4} \quad (1131)$$

$$\leq C \sum_{|a_1|+|a_2|=K-1} \left\|\frac{\partial_{a_1} \nabla v_\epsilon^b}{\Lambda^{|a_1|}}\right\|_{C^0} \left\|\frac{\partial_{a_2} \partial_b Q}{\Lambda^{|a_2|}}\right\|_{L^4} \quad (1132)$$

$$\leq C \Xi e_v^{1/2} E_M[Q]^{1/4}(t) \quad (1133)$$

$$\leq C\tau^{-1} E_M[Q]^{1/4}(t). \quad (1134)$$

Therefore, from (1129) I is bounded by

$$I \leq C\tau^{-1} E_M[Q](t). \quad (1135)$$

Now it remains to bound II in Line (1127). We estimate this term by

$$\int \left(\frac{\partial_a Q^{jl}}{\Lambda^{K-1}}\right)^3 \cdot \left(\Lambda^{-(K-1)} \partial_a \mathcal{R}^{jl}[\frac{\bar{D}U}{\partial t}]\right) dx \leq C\tau^{-1} \int \left|\frac{\partial_a Q^{jl}}{\Lambda^{K-1}}\right|^4 dx \quad (1136)$$

$$+ \int \tau^3 \Lambda^{-4(K-1)} \left|\partial_a \mathcal{R}^{jl}[\frac{\bar{D}U}{\partial t}]\right|^4 dx \quad (1137)$$

$$\leq C\tau^{-1} \left(E_M[Q](t) + \tau^4 \Lambda^{-4(K-1)} \int \left|\nabla^{K-1}[\frac{\bar{D}U}{\partial t}]\right|^4 dx\right) \quad (1138)$$

$$\leq C\tau^{-1}(E_M[Q](t) + A^4) \quad (1139)$$

which establishes (1116) and concludes the proof of (1114).

27.3 Material Derivative Estimates for the Transport-Elliptic Equation

With the estimates on spatial derivatives of the solution to (1094) in hand, we can now give bounds for the coarse scale material derivative $\frac{\bar{D}Q}{\partial t}$ of the solution to (1094) as well as its spatial derivatives.

We start by recalling the Transport-Elliptic equation

$$\frac{\bar{D}Q^{jl}}{\partial t} = \mathcal{R}^{jl}[\partial_i v_\epsilon^b \partial_b Q^{ik} + \frac{\bar{D}U^k}{\partial t}] \quad (1140)$$

181

and estimating each term by

$$\|\nabla^k \mathcal{R}^{jl}[\partial_i v_\epsilon^b \partial_b Q^{ik}]\|_{L^4} \leq C\|\nabla^{|k|-1}[\partial_i v_\epsilon^b \partial_b Q^{ik}]\|_{L^4} \tag{1141}$$

$$\leq C \sum_{|a_1|+|a_2|=|k|-1} \|\nabla^{|a_1|+1} v_\epsilon^b\|_{C^0} \|\nabla^{|a_2|+1} Q^{ik}\|_{L^4} \tag{1142}$$

$$\leq C \sum_{|a_1|+|a_2|=|k|-1} \Xi e_v^{1/2} \Lambda^{|a_1|} (\Lambda^{|a_2|} A e^{C\tau^{-1}|t-t(I)|}) \tag{1143}$$

$$\leq C\tau^{-1} \Lambda^{|k|-1} A e^{C\tau^{-1}|t-t(I)|} \tag{1144}$$

and

$$\left\|\nabla^k \mathcal{R}^{jl}[\frac{\bar{D}U}{\partial t}]\right\|_{L^4} \leq C \left\|\nabla^{|k|-1} \frac{\bar{D}U}{\partial t}\right\|_{L^4} \tag{1145}$$

$$\leq C\tau^{-1} \Lambda^{|k|-1} A. \tag{1146}$$

27.4   Cutting Off the Solution to the Transport-Elliptic Equation

We now use the preceding estimates on the solution to the Transport-Elliptic equation in order to establish (27.1).

Let Q_* be the solution to the Transport-Elliptic Equation (1094). Then Q_* satisfies the elliptic equation

$$\partial_j Q_*^{jl} = U^l,$$

but Q_* will not in general satisfy the properties desired by Theorem (27.1) because Q_* will not in general be compactly supported in time, and the estimates on its derivatives grow exponentially in time.

In order to fix this problem, we can simply cutoff Q_*. Namely, let $\bar{\eta}(t)$ be a cutoff function which is equal to $\bar{\eta}(t) = 1$ on the time interval $[t(I) - \tau, t(I) + \tau]$ containing the support of U^l, and such that $\bar{\eta}(t)$ is supported in $[t(I) - 3\tau/2, t(I) + 3\tau/2]$. Now define

$$Q^{jl} = \bar{\eta}(t) Q_*. \tag{1147}$$

Then Q solves $\partial_j Q^{jl} = U^l$, Q has integral 0, and Q satisfies all of the estimates stated in Theorem (27.1) because the cutoff ensures that the exponential factors of $e^{C\tau^{-1}|t-t(I)|}$ in the estimates for Q_* are all bounded. This concludes the proof of Theorem (27.1).   □

182

Acknowledgments

The author is endebted to his advisor Sergiu Klainerman for his guidance, for introducing him to the work of De Lellis and Székelyhidi and for introducing him to many great opportunities during his graduate studies. The author is thankful to professors C. Fefferman, A. Ionescu, V. Vicol and P. Constantin, and to his colleagues S.-J. Oh and J. Luk for all of their help. The author also thanks S.-J. Oh for suggestions regarding the proofs in Section (27), and thanks Vlad Vicol and Ting Zhang for corrections to earlier drafts of the book. The author is deeply grateful to C. De Lellis and L. Székelyhidi for discussions in 2011 about basic principles of convex integration and difficulties related to producing continuous Euler flows that helped to start the author's investigations in the subject. The author also thanks M. Gromov and C. Villani for insightful conversations about convex integration.

This work was supported by the NSF Graduate Research Fellowship Grant DGE-1148900 and the NSF Postdoctoral Research Fellowship DMS-1402370.

Appendices

Here we prove some facts about regular dodecahedra that have been used in the proof.

To prove the fact stated in Lemma (7.1), we establish the following.

Lemma .2. *Let $u \in \mathbb{R}^3$ be such that*

$$u \times f \neq 0 \tag{1148}$$

for all $f \in F$.

For every integer $n \geq 1$ there exists a constant $c > 0$ and a set of rotations O_m of the form

$$O_m = e^{\theta_m u \times} \qquad m = 1, \ldots, n \tag{1149}$$

with the property that

$$|f \circ O_m + f' \circ O_{m'}| \geq c$$
$$f, f' \in F \tag{1150}$$
$$m, m' - 1, \ldots, n$$

holds unless $f' - -f$ and $m' - m$.

Proof. We proceed by induction on n.

For $n - 1$, we take $\theta_1 = 0$, so that

$$O_1 = e^{0 u \times} = \text{Id}$$

is the identity. Then the property (1150) holds with $c = \min\{|f + f'| \mid f, f' \in F, f \neq -f'\}$.

The rest of our rotations will all be chosen from the one parameter group of rotations

$$O_\theta v = e^{\theta u \times} v. \tag{1151}$$

The fact that such maps are rotations can be seen by defining an auxiliary function

$$\phi(\theta) = <e^{\theta u \times} v, e^{\theta u \times} w>$$

for any $v, w \in \mathbb{R}^3$. The function $\phi(\theta)$ then satisfies

$$\phi(0) = <v, w>$$
$$\phi'(\theta) = <u \times (e^{\theta u \times} v), e^{\theta u \times} w> + <e^{\theta u \times} v, u \times (e^{\theta u \times} w)>$$
$$= 0$$

using the basic antisymmetry property of the cross product. The special case $\theta = 0$ corresponds to our initial rotation, the identity map.

Now assume by induction that we have rotations $e^{\theta_m u\times}$, $m = 1, \ldots, n$ for which the condition (1150) holds. To find another rotation with the property (1150), we observe that for $O_\theta = e^{\theta u\times}$, we have

$$|f \circ O_\theta + f' \circ O_m|^2 = |f \circ O_\theta|^2 + |f' \circ O_m|^2 + 2<e^{\theta u\times}f, e^{\theta_m u\times}f'>$$

$$= 2(1 + <e^{\theta u\times}f, e^{\theta_m u\times}f'>) \tag{1152}$$

$$= 2(1 + <e^{(\theta-\theta_m)u\times}f, f'>). \tag{1153}$$

By Cauchy-Schwartz, this quantity can be zero only when the function

$$\psi_{m,f,f'}(\theta) = \psi(\theta) = <e^{(\theta-\theta_m)u\times}f, f'> \tag{1154}$$

reaches a minimum value equal to -1.

We study $\psi(\theta)$ using the differential equation

$$\psi'''(\theta) + |u|^2\psi'(\theta)$$
$$= <(u\times)^3 e^{(\theta-\theta_m)u\times}f, f'> + |u|^2 <(u\times)e^{(\theta-\theta_m)u\times}f, f'> \tag{1155}$$

$$\psi'''(\theta) + |u|^2\psi'(\theta) = 0. \tag{1156}$$

This equation implies that ψ has the form

$$\psi(\theta) = A + B\cos(|u|\theta) + C\sin(|u|\theta). \tag{1157}$$

We must now argue that at least one of the coefficients B and C is nonzero to make sure that $\psi(\theta)$ takes on other values besides -1.

Our strategy is to use the differential equation (1156) to directly obtain a bound on the measure of the set

$$|\{\theta \in \mathbb{R}/(\frac{2\pi}{|u|}\mathbb{Z}) \mid |\psi(\theta) + 1| \leq c\}| \leq Ac^{1/2} \tag{1158}$$

for some constant A. This bound will follow from the conservation law

$$(\psi''(\theta))^2 + |u|^2(\psi'(\theta))^2 = (\psi''(\theta_m))^2 + |u|^2(\psi'(\theta_m))^2 \tag{1159}$$

once the conserved quantity has been shown to be nonzero at the initial $\theta = \theta_m$.

To see that both sides of (1159) are nonzero, let us assume by contradiction that both terms in (1159) are 0 at $\theta = \theta_m$. Namely,

$$\psi'(\theta_m) = <u \times f, f'> = 0 \tag{1160}$$

$$\psi''(\theta_m) = <(u\times)^2 f, f'> = 0 \tag{1161}$$

$$-<u \times f, u \times f'> = 0. \tag{1162}$$

By the basic properties of the cross product, we also know that

$$<u \times f, u> = 0. \tag{1163}$$

186

The condition (1148) implies that the vectors

$$\{u, f', u \times f'\}$$

form a basis of $\mathbb{R}^3$, so from (1160), (1162), and (1163) we conclude that

$$u \times f = 0 \tag{1164}$$

which contradicts (1148). It follows that the conserved quantity

$$(\psi''(\theta))^2 + |u|^2(\psi'(\theta))^2 = E^2 > 0 \tag{1165}$$

is strictly positive. The positivity of (1165) now implies the bound (1158), since (1159) implies that at least one of

- $|\psi'(\theta)| \geq \frac{E}{\sqrt{2}|u|}$

- or $|\psi''(\theta)| \geq \frac{E}{\sqrt{2}}$

holds at every point $\theta \in \mathbb{R}/(\frac{2\pi}{|u|}\mathbb{Z})$. Using (1158), we can find a new constant c' and a point $\theta_{m+1} \in \mathbb{R}/(\frac{2\pi}{|u|}\mathbb{Z})$ such that

$$|\psi_{m,f,f'}(\theta_{m+1}) + 1| \geq c' \qquad \text{for all } f, f' \in F, m = 1, \ldots, n.$$

This bound allows us to conclude (1150), and hence conclude our proof of Lemma .2. $\quad\square$

We now give a proof of the identity (120) which we restate here in the form

$$\delta^{jl} = \frac{1}{2} \sum_{f \in \mathbb{F}} f^j f^l \tag{1166}$$

$$= \frac{1}{4} \sum_{f \in F} f^j f^l. \tag{1167}$$

Proof of identity (120). To begin the proof, first observe that the bilinear form $G^{jl} = \sum_{f \in \mathbb{F}} f^j f^l = \frac{1}{2} \sum_{f \in F} f^j f^l$ is invariant under the action of the icosahedral group in the sense that for every symmetry $g : \mathbb{R}^3 \to \mathbb{R}^3$ of the dodecahedron, we have

$$\mathrm{Sym}^2 g(G)(u, w) = G(u \circ g, w \circ g) = G(u, w) \qquad \text{for all } u, w \in (\mathbb{R}^3)^*.$$

In particular, G is invariant under the subgroup A_5 of orientation-preserving symmetries of the dodecahedron. We claim that every $G^{jl} \in \mathcal{S}$ that is invariant under A_5 has the form

$$G^{jl} = C\delta^{jl}. \tag{1168}$$

Taking the trace of (1168) we must have $C = \frac{G^{jl}\delta_{jl}}{3}$. Setting $G^{jl} = \frac{1}{2}\sum_{f\in\mathbb{F}} f^j f^l$, we obtain the identity (120), which we have restated here as identity (1166).

In the language of representation theory, our claim is that the dimension of the space

$$\mathcal{S}^{A_5} = \{G^{jl} \mid \text{Sym}^2 g(G^{jl}) = G^{jl} \; \forall \; g \in A_5\}$$

of A_5-invariant elements of $\mathcal{S}$ is 1, or that the trivial representation of A_5 has multiplicity 1 in the representation $\mathcal{S} = \text{Sym}^2(\mathbb{R}^3)$. We now establish this claim using elementary character theory. To start, we recall the conjugacy classes in A_5:

- The identity class $c = \langle \text{Id} \rangle$ has 1 element

- The class $c = \langle (12)(34) \rangle$ has 15 elements of order 2, which rotate the dodecahedron by an angle π about an axis drawn between the midpoints of opposite edges

- The class $c = \langle (123) \rangle$ has 20 three-cycles, which rotate the dodecahedron by an angle $2\pi/3$ about an axis drawn between two opposite vertices

- The class $c = \langle (12345) \rangle$ has 12 five-cycles, which rotate the dodecahedron by an angle $2\pi/5$ about an axis drawn between the centers of opposite pentagonal faces

- The class $c = \langle (13524) \rangle = \langle (12345)^2 \rangle$ has 12 five-cycles, which rotate the dodecahedron by an angle $4\pi/5$ about an axis drawn between the centers of opposite pentagonal faces

Having a geometric description of how each element in A_5 acts on $\mathbb{R}^3$ allows us to compute the eigenvalues, and in particular the traces, of the corresponding matrices. These quantities are useful to compute because the dimension of $\mathcal{S}^{A_5}$ is equal to the trace of the operator

$$\frac{1}{|A_5|} \sum_{g\in A_5} \text{Sym}^2 g : \mathcal{S} \to \mathcal{S}$$

which projects to the invariant subspace $\mathcal{S}^{A_5}$ of $\mathcal{S}$. That is,

$$\dim \mathcal{S}^{A_5} = \frac{1}{|A_5|} \left(\sum_{g\in A_5} \text{tr Sym}^2 g \right). \tag{1169}$$

We can compute this trace using the plethysm formula

$$\text{tr Sym}^2(g) = \frac{1}{2} \left(\text{tr}\,(g^2) + \text{tr}^2(g) \right). \tag{1170}$$

In terms of the eigenvalues of the operators $g \in A_5$, the identity (1170) expresses the equality

$$\sum_{1 \le i \le j \le 3} \lambda_i \lambda_j = \frac{1}{2} \sum_{i=1}^{3} \lambda_i^2 + \frac{1}{2} \left(\sum_{i=1}^{3} \lambda_i \right)^2 \qquad \lambda_i \in \mathbb{C}. \qquad (1171)$$

For instance, when g is the identity element, there are three eigenvalues, all equal to 1, and we confirm that the dimension of $\mathcal{S}$ is $\frac{3+9}{2} = 6$; in higher dimensions, the above numerical identity generalizes to a very familiar formula for triangular numbers. The spectral interpretation of (1170) can be justified by noting that all the operators g we consider can be diagonalized on the complex vector space $\mathbb{C}^3 = \mathbb{C} \otimes \mathbb{R}^3$ since they satisfy a polynomial equation $g^k - 1 = 0$ with no repeated roots. Using symmetric products of these eigenvectors to form a basis for $\mathbb{C} \otimes \mathcal{S}$, one obtains (1170) from (1171).

All of our operators g are rotations in $\mathbb{R}^3$ by some angle θ; the trace of such an operator is given by $1 + e^{i\theta} + e^{-i\theta}$. For example, letting $\omega_k = e^{2\pi i/k}$, the three cycle acts with trace tr $(123) = 1 + \omega_3 + \omega_3^{-1} = 1 + \omega_3 + \omega_3^2 = (1 - \omega_3^3)/(1 - \omega_3) = 0$. Using these observations and equation (1170), we can build a table of conjugacy classes:

	Id	$\langle\langle(12)(34)\rangle\rangle$	$\langle\langle(123)\rangle\rangle$	$\langle\langle(12345)\rangle\rangle$	$\langle\langle(13524)\rangle\rangle$
#	1	15	20	12	12
tr	3	-1	0	$1 + \omega_5 + \omega_5^4$	$1 + \omega_5^2 + \omega_5^3$
tr Sym^2	6	2	0	γ_1	γ_2

Here $\gamma_1 = \frac{(1+\omega_5+\omega_5^4)^2 + (1+\omega_5^2+\omega_5^3)}{2}$ and $\gamma_2 = \frac{(1+\omega_5^2+\omega_5^3)^2 + (1+\omega_5+\omega_5^4)}{2}$.

When we sum the terms in the formula (1169) that come from the 5-cycles, there is a cancellation from $1 + \omega_5 + \omega_5^2 + \omega_5^3 + \omega_5^4 = 0$. Using this observation and some observations about the geometry of a regular pentagon, we can now see that

$$\dim \mathcal{S}^{A_5} = \frac{1}{60} \left(6 + 2 \cdot 15 + 12 \cdot \frac{1 + (1+\omega_5^2+\omega_5^3)^2 + (1+\omega_5+\omega_5^4)^2}{2} \right) \qquad (1172)$$

$$< \frac{1}{10} \left(1 + 5 + (1 + 2^2 + 3^2) \right) = 2. \qquad (1173)$$

Of course, the left-hand side is an integer, and it is at least 1 because $\delta^{jl} \in \mathcal{S}^{A_5}$ is nonzero, so this bound is enough to conclude the proof. $\qquad \square$

As a first application of the identity (1166), we calculate the angle between projectively distinct faces of the dodecahedron.

Lemma .3. *Suppose that* $x, y \in \mathbb{F}$ *and* $y \in \mathbb{F} \setminus \{x\}$, *then*

$$(x \cdot y)^2 = \frac{1}{5}. \qquad (1174)$$

Hence,

$$|x \wedge y|^2 = |x|^2|y|^2 - (x \cdot y)^2 = \frac{4}{5}. \tag{1175}$$

Proof. Let x be a fixed face of the projective dodecahedron. Let σ_x be a nontrivial rotation that fixes the pentagonal face x and permutes the 5 faces adjacent to x. Then σ_x has order 5, and cannot fix any of the five adjacent projective faces because it does not obtain the eigenvalue -1 while acting on $\mathbb{R}^3$, and the eigenvalue 1 is obtained only in the x direction itself since the rotation is nontrivial.

Hence, σ_x acts transitively on the other five faces in $\mathbb{F} \backslash \{x\}$, and therefore the number

$$(x \cdot y)^2 = (\sigma_x^k x \cdot \sigma_x^k y)^2 \tag{1176}$$

$$= (x \cdot \sigma_x^k y)^2 \tag{1177}$$

does not depend on the other face $y \in \mathbb{F} \backslash \{x\}$.

To calculate this number, we apply the identity (1166) to obtain

$$\delta^{jl} x_j x_l = \frac{1}{2} \sum_{f \in \mathbb{F}} f^j f^l x_j x_l \tag{1178}$$

$$|x|^2 = \frac{1}{2}(|x|^2 + 5(x \cdot y)^2) \tag{1179}$$

$$(x \cdot y)^2 = \frac{|x|^2}{5} = \frac{1}{5}. \tag{1180}$$

$\square$

As a final application of the identity (1166), we establish the linear independence Claim (1) of Lemma (7.3). Claim (2) of Lemma (7.3) can be proven using the same argument below.

Proof of Lemma (7.2). Suppose we have a linear relation

$$\sum_{f \in \mathbb{F}} \alpha_f f^j f^l = 0. \tag{1181}$$

As we discussed in the proof of (1174), every face $f_* \in \mathbb{F}$ has a set of five adjacent faces $\mathcal{N}(f_*)$. They can be obtained by starting with a single adjacent side x, and then rotating around the center of the face f_* using the five-cycle σ_{f_*}. These faces represent all five other faces of $\mathbb{F}$

$$\mathcal{N}(f_*) = \{\sigma_{f_*}^k x \mid k = 0, \ldots, 4\} = \mathbb{F} \backslash \{f_*\}.$$

To check this equality, note that σ_{f_*} only obtains an eigenvalue of ± 1 on f_* itself. Therefore σ_{f_*} must act transitively on the other five faces of $\mathbb{F}$ since σ_{f_*} has order five and has no fixed point on $\mathbb{F} \backslash \{f_*\}$.

By acting on the linear combination (1181) with the group $\langle \sigma_{f_*} \rangle$ we can produce potentially new relations. Averaging the relations obtained in this way gives

$$\alpha_{f_*} f_*^j f_*^l + \hat{\alpha}(f_*) \sum_{f \in \mathcal{N}(f_*)} f^j f^l = 0, \qquad (1182)$$

where $\hat{\alpha}(f_*) := \frac{1}{5} \sum_{f \in \mathcal{N}(f_*)} \alpha_f$.

From equation (1182), we now subtract both the summation over $\mathcal{N}(f_*)$ and the term $\hat{\alpha}(f_*) f_*^j f_*^l$. Recalling the identity (1166), we see that

$$(\alpha_{f_*} - \hat{\alpha}) f_*^j f_*^l = -\hat{\alpha} \sum_{f \subset \mathbb{F}} f^j f^l \qquad (1183)$$

$$(\alpha_{f_*} - \hat{\alpha}) f_*^j f_*^l = -2\hat{\alpha}\delta^{jl}. \qquad (1184)$$

Both sides of (1184) must be equal to zero, as otherwise they would differ in rank. It now follows that $\hat{\alpha}(f_*) = 0 = \alpha_{f_*}$ for all $f_* \in \mathbb{F}$, which concludes the proof of our lemma. $\qquad \square$

A   The Positive Direction of Onsager's Conjecture

In this Appendix, we provide a proof due to [CET94] of the positive direction of Onsager's conjecture. Formally, the proof proceeds by mollifying a weak solution to Euler in space $v \mapsto v_\epsilon = v * \phi_\epsilon$. The mollification does not commute with the nonlinearity and so v_ϵ becomes an "approximate solution" to Euler with a forcing term that involves the commutator $R_\epsilon^{j\ell} = (v^j v^\ell) * \phi_\epsilon - v_\epsilon^j v_\epsilon^\ell$. Multiplying the equation for v_ϵ by v_ϵ, integrating by parts and sending $\epsilon \to 0$, one obtains the result if the term involving R_ϵ is well-controlled. Establishing a **commutator estimate** for R_ϵ is the key ingredient in the proof. Our presentation of the details follows the treatment in [IO16], where the result is extended to compact Riemannian manifolds.

Let v be a weak solution to Euler of class $L_t^3 C_x^\alpha \cap C_t L_x^2 (I \times \mathbb{T}^n)$ on some open interval I for some $\alpha \in (1/3, 1)$. We want to show that the energy is constant in time. It suffices to show that

$$-\int_I \eta'(t) \int_{\mathbb{T}^n} \frac{|v|^2}{2}(t, x)dxdt = 0$$

for every smooth test function $\eta(t) \in C_c^\infty(I)$ with compact support in I. Using the fact that $e(t) = \int_{\mathbb{T}^n} \frac{|v|^2}{2}(t, x)dx$ is continuous in time, the fact that the energy is constant in t follows by taking η to approximate the characteristic function of any subinterval of I.

191

Let $\eta(t) \in C_c^\infty(I)$ be a smooth test function as above. Let $v_\epsilon(t, x) = v * \phi_\epsilon(t, x)$ denote a standard mollification of v in the spatial variable at length scale ϵ. We denote by $[v_\epsilon]_\delta$ a mollification of v_ϵ in the *time* variable at time scale δ. Using the fact that $v \in C_t L_x^2(I \times \mathbb{T}^n)$ we have

$$-\int_{I \times \mathbb{T}^n} \eta'(t) \frac{|v|^2}{2}(t, x) dx dt = -\lim_{\epsilon \to 0} \lim_{\delta \to 0} \int_{I \times \mathbb{T}^n} \eta'(t) \frac{|[v_\epsilon]_\delta|^2}{2}(t, x) dx dt.$$

(1185)

Using the self-adjointness of the mollification operators $*\phi_\epsilon$ and $[\ \cdot\ ]_\delta$, one can check that

$$-\int_{I \times \mathbb{T}^n} \eta'(t) \frac{|[v_\epsilon]_\delta|^2}{2}(t, x) dx dt = -\int_{I \times \mathbb{T}^n} v \cdot \partial_t [\eta(t)([v_\epsilon]_\delta * \phi_\epsilon)]_\delta dx dt,$$

where we use the Einstein summation convention to sum over ℓ. Now observe that $[\eta(t)([v_\epsilon]_\delta * \phi_\epsilon)]_\delta \in C_c^\infty(I \times \mathbb{T}^n)$ is a valid test function. Using the definition of weak solution to the Euler equations we then have

$$-\int_{I \times \mathbb{T}^n} \eta'(t) \frac{|[v_\epsilon]_\delta|^2}{2}(t, x) dx dt$$

$$= \int_{I \times \mathbb{T}^n} v^j v^\ell \nabla_j [\eta(t)([(v_\ell)_\epsilon]_\delta * \phi_\epsilon)]_\delta dx dt$$

$$+ \int_{I \times \mathbb{T}^n} p \nabla^\ell [\eta(t)([(v_\ell)_\epsilon]_\delta * \phi_\epsilon)]_\delta dx dt,$$

where we use the Einstein summation convention to abbreviate the sums over j and ℓ.

The rightmost term is 0 since the vector field v is divergence free, and so are its mollifications. Also, using standard regularization properties of mollification and the fact that $v \in C_t L_x^2(I \times \mathbb{T}^n)$, we can take $\delta \to 0$ at this point and the first term converges to

$$\int_{I \times \mathbb{T}^n} v^j v^\ell \nabla_j \eta(t)((v_\ell)_\epsilon * \phi_\epsilon) dx dt = \int_{I \times \mathbb{T}^n} \eta(t)(v^j v^\ell) * \phi_\epsilon \nabla_j (v_\ell)_\epsilon dx dt.$$

This expression is similar to the expression

$$\int_{I \times \mathbb{T}^n} \eta(t) v_\epsilon^j v_\epsilon^\ell \nabla_j (v_\ell)_\epsilon dx dt = \frac{1}{2} \int_{I \times \mathbb{T}^n} \eta(t) v_\epsilon^j \nabla_j |v_\epsilon|^2 (t, x) dx dt = 0,$$

which vanishes because v_ϵ is divergence free. Subtracting the latter from the former we obtain

$$-\int_I \eta'(t) \int_{\mathbb{T}^n} \frac{|v|^2}{2}(t, x) dx dt = \lim_{\epsilon \to 0} \int_{I \times \mathbb{T}^n} \eta(t) R_\epsilon^{j\ell} \nabla_j (v_\epsilon)_\ell dx dt \qquad (1186)$$

$$R_\epsilon^{j\ell} = (v^j v^\ell) * \phi_\epsilon - v_\epsilon^j v_\epsilon^\ell.$$

A standard estimate for convolutions gives

$$\|\nabla_j(v_\epsilon)_\ell(t,x)\|_{C^0(\mathbb{T}^n)} \leq C\epsilon^{-1+\alpha}\|v(t,\cdot)\|_{C_x^\alpha}$$
$$\Rightarrow \|\nabla_j(v_\epsilon)_\ell(t,x)\|_{L_t^3 C_x^0(I\times\mathbb{T}^n)} \leq C\epsilon^{-1+\alpha}\|v\|_{L_t^3 C_x^\alpha}.$$

(1187)

The key estimate to conclude the proof is the following *commutator estimate* of [CET94]:

$$\|R_\epsilon(t,x)\|_{C^0(\mathbb{T}^n)} \leq C\epsilon^{2\alpha}\|v(t,\cdot)\|_{C_x^\alpha(\mathbb{T}^n)}^2$$
$$\Rightarrow \|R_\epsilon(t,x)\|_{L_t^{3/2}C_x^0(I\times\mathbb{T}^n)} \leq C\epsilon^{2\alpha}\|v(t,\cdot)\|_{L_t^3 C_x^\alpha(\mathbb{T}^n)}^2$$

(1188)

Combining the estimates (1187) and (1188), the right-hand side of (1186) has size bounded by a constant times $\epsilon^{-1+3\alpha}$, and hence vanishes as $\epsilon \to 0$ since $\alpha > 1/3$. (Here we use the fact that $\mathbb{T}^n$ has finite volume.)

It now remains to prove (1188). We use an expression for the commutator that was used to establish the time regularity theory in [Ise13a], which is slightly different from (smoother than) the one in [CET94].

To prove (1188), let us omit the variable t and write

$$R_\epsilon^{j\ell}(x) = \int_{\mathbb{R}^n} v^j(x+h)v^\ell(x+h)\phi_\epsilon(h)dh$$
$$- \int_{\mathbb{R}^n} v^j(x+h_1)\phi_\epsilon(h_1)dh_1 \int_{\mathbb{R}^n} v^\ell(x+h_2)\phi_\epsilon(h_2)dh_2.$$

This expression is similar to the formula for the variance of a random variable $\mathrm{Var}(V) = E[V^2] - (E[V])^2$ since $\int_{\mathbb{R}^n} \phi_\epsilon(h)dh = 1$. In the same way that a variance can be written quadratically as $\mathrm{Var}(V) = E[(V - E[V])^2]$, one can check (using $\int_{\mathbb{R}^n} \phi_\epsilon(h)dh = 1$) that

$$R_\epsilon^{j\ell}(x) = \int_{\mathbb{R}^n} (v^j(x+h) - v_\epsilon^j(x))(v^\ell(x+h) - v_\epsilon^\ell(x))\phi_\epsilon(h)dh.$$

To estimate R_ϵ, we decompose

$$v(x+h) - v_\epsilon(x) = (v(x+h) - v_\epsilon(x+h)) + (v_\epsilon(x+h) - v_\epsilon(x)),$$

and use the estimates

$$\|v - v_\epsilon\|_{C^0} \leq C\epsilon^\alpha\|v\|_{C_x^\alpha(\mathbb{T}^n)}$$
$$|v_\epsilon(x+h) - v_\epsilon(x)| \leq \|\nabla v_\epsilon\|_{C^0}|h| \leq C\epsilon^{-1+\alpha}\|v\|_{C_x^\alpha}|h|$$
$$\int_{\mathbb{R}^n} |h|^k \phi_\epsilon(h)dh \leq C\epsilon^k \text{ for } k = 0,1,2.$$

The resulting bound has the form (1188), which concludes the proof.

193

B Simplifications and Recent Developments

Since the initial appearance of this monograph, there have been many developments in our understanding of the method of convex integration and its applications to studying weak solutions to the Euler and related equations. In particular, the literature now contains several ways to simplify the proofs of the results in the monograph, and we feel it is useful to outline these simplifications here.

A much shorter proof of the existence of $(1/5 - \epsilon)$-Hölder solutions that fail to conserve energy is given in [BDLS13]. The proof is more closely based on the original constructions of [DLS13, DLS14] but incorporates several key ideas in this monograph; see [BDLIS15] for a self-contained presentation that includes a more detailed technical comparison to the present work. The most substantial simplification in [BDLS13] is a way to avoid the mollification along the flow technique by approximating the Reynolds stress during a small time interval by a solution to a free transport equation. The authors also approximate the nonlinear phase functions in the construction by their linear initial values. One must be cautious that these particular techniques introduce additional error terms that rely on the smallness of the time scale to control; it is possible that at higher regularity and longer time scales such terms may not be acceptable. For example, in the later work of [BDLS15], which demonstrates existence of energy nonconserving solutions with Onsager critical spatial regularity $v \in C^0_{t,x} \cap L^1_t C^{1/3-\epsilon}_x$, the authors rely on mollification along the flow to control an error term that lives on a relatively long time scale.

Several simplifications to the approach in the present paper may also be extracted from the works of [IOa, IV15]. Each of these works introduces an improvement that avoids the need for super-exponential growth of frequencies during the iteration (in particular, the proof in Section (11) is much simpler after removing the restriction $N \geq \Xi^\eta$ in Lemma (10.1)). In [IOa], the authors introduce a family of operators that give solutions to the equation $\partial_j Q^{jl} - U^l$ with compact support in a small ball; this technique combined with the use of localized waves allows one to terminate the parametrix expansion of Section (26) after only four terms and results in exponential growth of frequencies. In [IV15], the authors use frequency localization operators in combination with a "Microlocal Lemma" to build the oscillatory corrections; these techniques eliminate the need for a parametrix expansion entirely and simplify the estimates of Sections (26)–(27).

We conclude by listing some further recent developments related to the results proven here:

- In [BDLS13], it is shown that every smooth strictly positive function $e(t)$ on a closed interval can be realized as the energy profile of a solution in the class $v \in C^\alpha_{t,x}$ if $\alpha < 1/5$.

- In [BDLS15], it is shown that there exist nontrivial solutions with

Onsager-critical spatial regularity $v \in C_{t,x}^0 \cap L_t^1 C_x^{1/3-\epsilon}$ that have compact support in time. The proof builds on a key observation in [Buc15] that it is possible to modify the construction in [BDLS13] and incorporate localized estimates to restrict the consistent appearance of bad error terms to a set of times of measure zero.

- The paper [Ise13a] presents a time regularity theory for weak solutions assuming Hölder regularity in space $v \in L_t^\infty C_x^\alpha(I \times \mathbb{T}^n)$, $0 < \alpha < 1$. It is shown that many time regularity properties of solutions in the present work, such as the improved regularity of the material derivative, are consequences of the Euler equations rather than artifacts of the construction. Results include:

  - If $v \in L_t^\infty C_x^\alpha$, $0 < \alpha < 1$, then $v \in C_{t,x}^\alpha$ and $p \in C_{t,x}^{2\alpha-\epsilon}$ for any $\epsilon > 0$.

  - If $1/(1-\alpha)$ is not an integer, then every particle trajectory is of class $C^{1/(1-\alpha)}$ in time, and the velocity field is $C^{\alpha/(1-\alpha)}$ in time along each trajectory. Similarly, every advective derivative $\frac{D^r v}{\partial t^r}$ of order r is continuous in (t,x) for $r < \alpha/(1-\alpha)$.

  - If $0 < \alpha \leq 1/3$, the energy profile $e(t) - \frac{1}{2} \int_{\mathbb{T}^n} |v|^2(t,x)dx$ is of class $C_t^{2\alpha/(1-\alpha)}$ in time.

  It is conjectured in this work that the energy profile of a generic Euler flow at regularity below $1/3$ fails to be more regular than $C^{2\alpha/(1-\alpha)}$ on every open interval and in particular fails to exhibit energy dissipation on every open interval. See [IOa] for further discussion.

- In [IOa], the authors extend the results of the present manuscript to construct nonzero solutions on $I \times \mathbb{R}^3$ with compact support in space and time, or as compactly supported perturbations of smooth Euler flows. The authors show that any smooth divergence free vector field of compact support that conserves both linear and angular momentum is a weak-* limit in $L_{t,x}^\infty$ of a sequence of compactly supported $C_{t,x}^{1/5-\epsilon}$ solutions. It is also shown that the solutions constructed with this method fail to belong to the class $v(t,\cdot) \notin W_x^{1/5,1} \cup C_x^{1/5}(B)$ on every time slice and open ball B contained in the support of the iteration. This failure of higher regularity does not involve changing the construction and applies to the solutions of the present work and [BDLS13] as well.

- In [IOa, IOb], building on [BDLS13], the authors show that for $0 < \alpha < 1/5$ every non-negative function of the class $e(t) \in C_t^{2\alpha/(1-\alpha)+\epsilon}$ is the energy profile of a weak solution in the class $v \in C_{t,x}^\alpha$. The regularity of the energy profile proven in [Ise13a] is therefore sharp in this range.

- In [IV15], the authors extend convex integration to a class of active scalar equations, generalizing a nonuniqueness result for $L_{t,x}^\infty$ solutions in [Shv11]. These equations have the form

$$\partial_t \theta + \partial_j(\theta u^j) = 0$$

where $\theta : I \times \mathbb{T}^d \to \mathbb{R}$ is a scalar and the velocity u is divergence free and is given by $u = T[\theta]$, where T is a Fourier multiplier of order 0. Failure of energy conservation for α-Hölder solutions is expected for $\alpha < 1/3$ just as for Euler. The main result shows that, under a non-degeneracy condition on the multiplier, there exist nonzero compactly supported solutions for $0 < \alpha < \frac{1}{1+4d}$, such solutions can realize arbitrary smooth mean-0 initial data, and every smooth scalar field that conserves the integral can be approximated in $L_{t,x}^\infty$ weak-* by solutions in the above class. It is shown that the nondegeneracy condition on T is necessary for the above h-principle to hold; in particular, it is crucial that the multiplier cannot be an odd function of frequency, which excludes the important case of the surface quasigeostrophic equations.

- In [CS], an h-principle result for $L_{t,x}^\infty$ stationary solutions to incompressible Euler is proved. The work builds on the techniques of [DLS09, DLS10].

- In [DS16], the Cauchy problem for incompressible Euler on $\mathbb{T}^3$ is considered. It is shown that the set of wild initial data is dense in $L^2(\mathbb{T}^3)$, where an initial datum is called wild if it admits infinitely many "admissible" solutions of class $C_{t,x}^{1/5-\epsilon}$. An admissible solution here is one for which $\int_{\mathbb{T}^3} |v|^2(t,x)dx \leq \int_{\mathbb{T}^3} |v|^2(0,x)dx$ for all t; we note that the solutions of Theorem (0.2) are not admissible, since any admissible solution must be equal to the unique smooth solution for as long as the latter exists. This result extends work of [SW12] in the setting of $L_{t,x}^\infty$ solutions. The authors also prove an h-principle result that essentially characterizes those symmetric tensors R^{jl} that arise as the error $R^{jl}(t,x) = \bar{v}^j \bar{v}^l - \int (\tilde{v}^j \tilde{v}^l)d\mu(\tilde{v})$ in taking a weak limit of solutions to incompressible Euler as described in Section 1.

There are many important questions we still do not know how to answer concerning weak solutions to the Euler equations. Among these the most obvious, fundamental problems still unsolved are Onsager's conjecture and the question of whether uniqueness of solutions fails at regularity below C^1. It is our hope that the method of convex integration and the ideas of this monograph may continue to be useful in further advancing the study of these and related problems.

References

[BDLIS15] Tristan Buckmaster, Camillo De Lellis, Philip Isett, and László
           Székelyhidi, Jr. Anomalous dissipation for 1/5-Hölder Euler
           flows. *Ann. of Math. (2)*, 182(1):127–172, 2015.

[BDLS13]   T. Buckmaster, C. De Lellis, and L. Székelyhidi, Jr. Trans-
           porting microstructures and dissipative Euler flows, Preprint.
           arXiv preprint arXiv:arXiv:1302.2815v3, 2013.

[BDLS15]   Tristan Buckmaster, Camillo De Lellis, and László Székelyhidi,
           Jr. Dissipative Euler flows with Onsager-critical spatial reg-
           ularity. *Communications on Pure and Applied Mathematics*,
           2015.

[BHSM99]   F. Bethuel, G. Huisken, K. Steffen, and S. Müller. Variational
           models for microstructure and phase transitions. In *Calcu-
           lus of Variations and Geometric Evolution Problems*, volume
           1713 of *Lecture Notes in Mathematics*, pages 85–210. Springer,
           Berlin/Heidelberg, 1999.

[Buc15]   Tristan Buckmaster. Onsager's conjecture almost everywhere
           in time. *Comm. Math. Phys.*, 333(3):1175–1198, 2015.

[CDLS12a] Antoine Choffrut, Camillo De Lellis, and László Székelyhidi, Jr.
           Dissipative continuous Euler flows in two and three dimensions.
           arXiv preprint arXiv:1205.1226, 2012.

[CDLS12b] Sergio Conti, Camillo De Lellis, and László Székelyhidi, Jr. *h*-
           principle and rigidity for $C^{1,\alpha}$ isometric embeddings. In *Non-
           linear partial differential equations*, volume 7 of *Abel Symp.*,
           pages 83–116. Springer, Heidelberg, 2012.

[CET94]   P. Constantin, W. E, and E. Titi. Onsager's conjecture on the
           energy conservation of Euler's equation. *Comm. Math. Phys.*,
           165(1):207–209, 1994.

[Cho13]   Antoine Choffrut. *h*-principles for the incompressible Euler
           equations. *Arch. Ration. Mech. Anal.*, 210(1):133–163, 2013.

[CS]   A. Choffrut and L. Székelyhidi, Jr. Weak solutions to the sta-
           tionary incompressible Euler equations. To appear in. *SIAM
           Journal of Math. Anal.*

[DLSa]   C. De Lellis and L. Székelyhidi, Jr. Continuous dissipative Eu-
           ler flows and a conjecture of Onsager. To appear in. *Proceedings
           of the 6th European Congress of Mathematics.*

[DLSb]   C. De Lellis and L. Székelyhidi, Jr. The h-principle and the equations of fluid dynamics. To appear in. *Bull. of Amer. Math. Soc.*

[DLS09]   C. De Lellis and L. Székelyhidi, Jr. The Euler equations as a differential inclusion. *Ann. Math.*, 170(3):1417–1436, 2009.

[DLS10]   C. De Lellis and L. Székelyhidi, Jr. On admissibility criteria for weak solutions of the Euler equations. *Arch. Ration. Mech. Anal.*, 195(1):225–260, 2010.

[DLS13]   Camillo De Lellis and László Székelyhidi, Jr. Dissipative continuous Euler flows. *Invent. Math.*, 193(2):377–407, 2013.

[DLS14]   Camillo De Lellis and László Székelyhidi, Jr. Dissipative Euler flows and Onsager's conjecture. *J. Eur. Math. Soc. (JEMS)*, 16(7):1467–1505, 2014.

[DS16]   Sara Daneri and László Székelyhidi, Jr. Non-uniqueness and h-principle for Hölder-continuous weak solutions of the Euler equations. *arXiv preprint arXiv:1603.09714*, 2016.

[Eyi94]   G. L. Eyink. Energy dissipation without viscosity in ideal hydrodynamics. I. Fourier analysis and local energy transfer. *Phys. D.*, 78(3-4):222–240, 1994.

[Fri95]   U. Frisch. *Turbulence. The Legacy of A. N. Kolmogorov.* Cambridge University Press, Cambridge, 1995.

[Gro86]   M. Gromov. *Partial Differential Relations.* Springer-Verlag, 1986.

[IOa]   P. Isett and S.-J. Oh. On nonperiodic Euler flows with Hölder regularity. To appear in. *Arch. Rat. Mech. Anal.*

[IOb]   P. Isett and S.-J. Oh. On the kinetic energy profile of Hölder continuous Euler flows. To appear in. *Annales Inst. H. Poincaré Anal. Non Linéaire.*

[IO16]   Philip Isett and Sung-Jin Oh. A heat flow approach to Onsager's conjecture for the Euler equations on manifolds. *Trans. Amer. Math. Soc.*, 368(9):6519–6537, 2016.

[Ise13a]   P. Isett. Regularity in time along the coarse scale flow for the Euler equations. Preprint. 2013.

[Ise13b]   Philip Isett. *Hölder continuous Euler flows with compact support in time.* ProQuest LLC, Ann Arbor, MI, 2013. Thesis (Ph.D.), Princeton University.

REFERENCES

[IV15]   Philip Isett and Vlad Vicol. Hölder continuous solutions of active scalar equations. *Annals of PDE*, 1(1):1–77, 2015.

[Kir03]   B. Kirchheim. Rigidity and geometry of microstructures. Habilitation thesis, University of Leipzig. 2003.

[KMŠ03]   B. Kirchheim, S. Müller, and V. Šverák. *Studying Nonlinear PDE by geometry in matrix space*, pages 347–395. Springer-Verlag, 2003.

[Kol41]   A. N. Kolmogorov. The local structure of turbulence in an incompressible viscous fluid. *C. R. (Doklady) Acad. Sci. URSS (N.S.)*, 30:301–305, 1941.

[MŠ56]   S. Müller and V. Šverák. Convex integration for Lipschitz mappings and counterexamples to regularity. *Ann. Math.*, 63:20–63, 1956.

[Nas54]   J. Nash. C^1 isometric embeddings i, ii. *Ann. Math.*, 60:383–396, 1954.

[Ons49]   L. Onsager. Statistical hydrodynamics. *Nuovo Cimento (9)*, 6 Supplemento(2 (Convegno Internazionale di Meccanica Statistica)):279–287, 1949.

[Sch93]   V. Scheffer. An inviscid flow with compact support in space-time. *J. Geom. Anal.*, 3(4):343–401, 1993.

[Shn97]   A. Shnirelman. On the nonuniqueness of weak solution of the Euler equation. *Comm. Pure Appl. Math.*, 50(12):1261–1286, 1997.

[Shn00]   A. Shnirelman. Weak solutions with decreasing energy of incompressible Euler equations. *Comm. Math. Phys.*, 210(3):541–603, 2000.

[Shv10]   R. Shvydkoy. Lectures on the Onsager conjecture. *Discrete Contin. Dyn. Syst. Ser. S 3*, 3:473–496, 2010.

[Shv11]   R. Shvydkoy. Convex integration for a class of active scalar equations. *J. Amer. Math. Soc.*, 24(4):1159–1174, 2011.

[SW12]   László Székelyhidi, Jr. and Emil Wiedemann. Young measures generated by ideal incompressible fluid flows. *Arch. Ration. Mech. Anal.*, 206(1):333–366, 2012.

Index